INTERNATIONAL CENTRE FOR MECHANICAL SCIENCES

COURSES AND LECTURES - No. 25

WITOLD NOWACKI
UNIVERSITY OF WARSAW

THEORY OF MICROPOLAR ELASTICITY

COURSE HELD AT THE DEPARTMENT
FOR MECHANICS OF DEFORMABLE BODIES
JULY 1970

UDINE 1970

SPRINGER-VERLAG WIEN GMBH

Originally published by Springer-Verlag Wien-New York in 1972

ISBN 978-3-211-81078-1 ISBN 978-3-7091-2720-9 (eBook)
DOI 10.1007/978-3-7091-2720-9

PREFACE

The purpose of this book is to give an account of the theory of elastic continuum with oriented particles. The fundamentals of the theory were devised in 1909 by the brothers Eugène and François Cosserat. This theory, in spite of the novelty of the comprised idea, was underrated during the authors'life, and unnoticed for many years thereafter. Only in the last decade has the theory been developed, and finds its applications mainly in the problems of the deformation of granular media and multi-molecular systems.

In the present lectures the assumptions, fundamental relations, and the equations of the micropolar theory of elasticity are discussed. The theory is presented on the basis of the classical theory of elasticity. An account is given of the statical and dynamical problems of the micropolar theory of elasticity and a great portion is devoted to the micropolar thermoelasticity.

I take the liberty of expressing my gratitude to Professor Luigi Sobrero, secretary General of the Centre International des Sciences Mécaniques, who encouraged me to prepare these lectures.

W. Nowacki

Udine, July 1970.

Chapter 1
General Theorems of Linear Micropolar Elasticity.

1. 1 Introduction

The classical theory of elasticity is based on an ideal model of an elastic, continuous medium in which the loadings are transmitted through an area element dA in the body by means of the stress vector only.

Taking into account this assumption, we obtain the description of the body-deformation by means of the symmetric tensors of deformation γ_{ji} and stress σ_{ji} .

The results obtained with the application of the classical theory of elasticity are in harmony with experiments carried out on many construction materials (various sorts of steel, aluminium and concrete) provided the stresses are within the limits of elasticity of the material. However, in many cases, remarkable discrepancies between theory and experiments have been observed. This refers, first of all, to such states of stress that show considerable stress gradients. The stress concentrations in the neighborhood of holes, notches and cracks, may serve, as examples of such states.

The discrepancy between the classical theory of elasticity and the experiments is particularly striking in dynamical problems, as in the case of elastic vibrations characterized by high frequency and small wave-lengths, i.e. for ultrasonic waves. This discrepancy results from the fact that for high frequencies and small wave-lengths, the influence of the body microstructure becomes significant.

The classical theory of elasticity eventually fails in the case of vibrations of grain bodies and multimolecular bodies such as polymers. The influence of the microstructure becomes here considerable, and results in development of new types of waves, unknown in the classical theory of elasticity. W.Voigt+) tried to correct these deficiences of the classical theory of elasticity by taking into account the assumption that the interaction between two parts of the body through an area element dA (inside the body) is transmitted not only by the force vector $\underline{p}\,dA$, but also by the moment vector $\underline{m}\,dA$. Such an assumption relays the fact that not only the force stresses σ_{ji} but also the couple stresses μ_{ji} act on the faces of an elementary parallelepiped. From the equilibrium equations for the parallelepiped it results that the stresses σ_{ji}, μ_{ji} are asymmetric.

The complete theory of asymmetric elasticity was developed in 1909 by the brothers E. and F. Cosserat++). In

---

+) Voigt, W.: Theoretische Studien über die Elastizitätsverhältnisse der Kristalle. Abh. Ges. Wiss. Göttingen, 34 (1887)

++) Cosserat, E., and F. Cosserat: Théorie des corps déformables. A. Hermann, Paris, 1909.

their theory, which from the very beginning was non-linear, the brothers E. and F. Cosserat made the next step farther. They assigned to each molecule a perfectly rigid trihedron, which, during the precess of deformation, underwent not only the displacement but also the rotation. Thus, was created such an elastic medium that its points obtained an orientation and of which we speak of as "rotation of a point".

In the Cosserat-theory of elasticity, the deformation of the body is described by a displacement vector $\underline{u}(\underline{x},t)$ and an independent rotation vector $\underline{\varphi}(\underline{x},t)$. The assumption that a medium consisted of material elements of six degrees of freedom led to the consequence of the asymmetry of strain and stress tensors.

In spite of the novelity of the idea, the work of the brothers E. and F. Cosserat was not duly appreciated during their lifetime and was unnoticed for a good while. Perhaps this was so because the theory was non-linear from the beginning, the notation was unclear and the theory included a number of problems departing from the frames of the theory of elasticity. The authors considered the problems of an ideal liquid, of a quasi-elastic medium "Mac Cullagh and Kelvin ether". They attempted to synthetize the Poincaré and Lorentz theory of the dynamics of electrons. The work of the Cosserat brothers can be treated as an attempt to create the unified field theory, containing mechanics, optics and electrodynamics, and combined with a general principle of the least action.

The Cosserat brothers did not derive, in their theory, the constitutive equations. They were interested in obtaining the integrand (the action density) in the general variational problem of Hamilton. The principle of the least action in the Cosserats' formulation for any continuum represents a scalar functional with geometric, kinematic and kinetic variables.

E. and F. Cosserat postulated the invariance of the action density with respect to the group of Euclidean displacements. The obtained group of transformations is a seven-parameter one. The requirement of invariance is, here, equivalent to the existence of the conservation principles: namely, the conservation of the mechanical energy, conservation of momentum, and conservation of angular momentum.

The research in the field of general theories of continuous media - conducted in the last fifteen years - drew attention of scientists to Cosserats'work. Now we observe a revival of interest in this theory. The next works on asymmetric elasticity belong to a period in recent years. At first, the interest of the research workers was concentrated on the simplified Cosserat theory, namely the theory of asymmetric elasticity of the so-called Cosserat pseudo-continuum. By this name, we understand, here, a continuum for which asymmetric stresses σ_{ji}, μ_{ji} may occur, while the displacements of a body are described by a single displacement vector $\underline{u}(\underline{x},t)$ only. This simplified model was already considered by the Cosserats, who called it the case with the latent trihedron.

Among the papers on the Cosserat pseudo-continuum, there are the works by C. Truesdell and R.A. Toupin*), which deserve our attention. These papers refer to linear and non-linear theory of elasticity in such a continuum. The theory of the Cosserat pseudo-continuum was developed by G. Grioli**), R.D. Mindlin and H.F. Tiersten***). In particular, the last authors obtained a series of interesting results in the linear theory, by introducing the potentials and stress functions, and discussing the problems of the elastic waves. Somewhat later, the theory of asymmetric elasticity, in the Cosserats'sense, was developed. Here, we should mention the paper by R.A. Toupin+), referring to the non-linear theory of the asymmetric elasticity. An interesting presentation of the theory was given by A.C. Eringen and E.S. Suhubi++).

---

*) Truesdell, C., and R.A. Toupin: The classical field theories. Encyclopedia of Physics, 3, No 1, Springer Verlag, Berlin, 1960.

**) Grioli, G.: Elasticità asimmetrica. Ann. di Mat. pura ed appl. Ser. IV. 50, (1960).

***) Mindlin, R.D. and H.F. Tiersten: Effects of couple-stresses in linear elasticity. Arch. Mech. Analysis, 11, (1962), 385.

+) Toupin, R.A.: Theories of elasticity with couple-stresses. Arch. Mech. Analysis. 17, (1964), 85.

++) Eringen, A.C., and E.S. Suhubi: Non linear theory of micro elastic solids. Int. J. Eng. Sci. 2 (1964), 189, 389.

The fundamentals of the linear theory of the general Cosserat continuum have been developed by W. Günther *) and H. Schaefer **). The first of the authors discussed in detail the one - two - and three-dimensional Cosserat model of continuum and pointed out the significance of the Cosserat theory in the dislocation problems. The latter of these authors reconstructed the fundamentals of the Cosserat theory for the problems of the plane state of strain. The general relations and the equations of the linear micropolar elasticity were given by E.V. Kuvshinskii, A.L. Aero ***) and by N.A. Palmov +). The fundamentals of thermoelasticity within the frames of the Cosserat medium were given by W. Nowacki ++).

A comprehensive problem article on asymmetric elasticity, together with the extensive bibliography of the

---

*) Günther, W.: Zur Statik und Kinematik des Cosseratschen Kontinuum. Abh. Braunschweig. Wiss. Ges 10, (1958), 85.
**) Schaefer, H.: Versuch einer Elastizitätstheorie des zweidimensionalen Cosserat-Kontinuum, Misz. Angew. Math. Festschrift Tollmien, Berlin, 1962, Akademie Verlag.
***) Kuvshinskii, E.V., and A.L. Aero: Continuum theory of asymmetric elasticity, (in Russian). Fizika Tverdogo Tela 5, (1963), 2592.
+) Palmov, N.A.: Fundamental equations of the theory of asymmetric elasticity (in Russian). Prikl. Mat. Mekh. 28 (1964), 401.
++) Nowacki W.: Couple-stresses in the theory of thermoelasticity. Proc. of the IUTAM symposia, Vienna, June 22 - 28 1966. Springer Verlag, Wien, 1968.

works on the subject, can be found in the study of H. Schaefer[+)].

The present monograph is concerned with the general linear theory of Cosserats'medium. However we impose certain restrictions on the micropolar Cosserat medium. We confine ourselves to the problems of elastic, homogeneous, isotropic and centrosymmetric bodies. These notions are well known from the classical theory of elasticity. Let us recall that the elastic deformation is defined as the process possessing the property of recovering the shape of the body when the forces producing the deformations are removed. By the homogeneity of the body, we understand that the density ρ, the material-constants and the rotational inertia $\mathcal{J}$ are independent of the position. The isotropy denotes the independence of the elastic properties of the direction. The invariance of the elastic properties of the material (invariance of the elastic coefficients), with respect to the inversion of the system of coordinates, is called the centro-symmetry or the symmetry with respect to the centre.

A state in which the body is undeformed and unstressed, with no external effects, is defined as the natural state of the body. The absolute temperature of such a state is denoted by T_0 = const. The external loadings and the heating of the body undergo the displacement $\underline{u}$ and rotation $\underline{\varphi}$, and the

+) Schaefer, H.: Das Cosserat-Kontinuum. Z.A.M.M. 47 (1967), 485

temperature changes by $\theta = T - T_0$. Here, $T(\underline{x},t)$ is the absolute temperature, a function of time and position.

It is assumed that the change of temperature accompanying the strain does not induce any essential change in the values of material and thermal coefficients. These coefficients are regarded as independent of T.

Here, like in the classical, linear theory of elasticity, we assume that the deformations are small, and the squares and products of the deformations are negligible with respect to the linear terms. It is also assumed that the relations between the state of strain and stress are linear, and the increments of temperature are inconsiderable, i.e. $\left|\frac{\theta}{T}\right| \ll 1$.

In this monograph, we shall refer many times to the ideas and results of the classical theory of elasticity which constitutes one of particular cases of Cosserats' theory here considered.

In this book, the stress will be laid on the presentation of the general linear theory of the Cosserat continuum. This theory does not have yet the complete experimental verification. We know, merely, the order of magnitudes and the mutual relations between the six material constants. Here, we encounter a striking example in which the theory is ahead of the experiment. However, the complete correspondence of the experiment and the theory exists in the case of discrete media (spatial grillages) where all the material constants can be uniquely determined. When passing from such a grillage to a continuous medium, we obtain exactly Cosserats'continuum +), ++).

1. 2 Equations of Equilibrium and Motion.

Let us consider an elastic body occupying the region B of volume V and bounded by the surface A. Let the body be deformed under the influence of external actions, consisting - in addition to the external body - and surface - loads (forces and couples) - of temperature variations, chemical reactions, electromagnetic influences, etc. Let $\underline{X}$ denote the body force per unit volume, and $\underline{Y}$ - the body couple. Both the body forces and body couples are functions of position $\underline{x}$ and time t.

Surface force (or surface traction) $\hat{\underline{p}}\,dA$ is defined as the force acting on the infinitesimal surface element dA. Intensity $\hat{\underline{p}}$ is the finite vector applied to an arbitrary point of the surface element dA. $\hat{\underline{m}}\,dA$ denotes the couple acting on element dA of the surface A. Quantities $\hat{\underline{p}}$ and $\hat{\underline{m}}$ are functions of position $\underline{x}$ and time t.

In the process of deformation (the distribution of particles constituting the body changes), certain internal

---

+) Kaliski, S.: On a model of the continuum with an essentially non-symmetric tensor of mechanical stress. Arch. Mech. Stos. 15, 1 (1963), 33.

++) Askar, A. and A.S. Cakmak: A structural model of a micropolar continuum. Int. J. Eng. Sci. 6, 10 (1968), 583.

forces appear in the body counteracting its deformation and tending to bring the body back to the original state of equilibrium. These forces are called stresses; they result from the particle interactions, short range forces characterized by a very small radius of action. The influence of these forces is reduced to the range of intermolecular distances.

The theory of elasticity is a macroscopic theory, and the distances considered within its frames are much greater than the intermolecular distances; hence, it can be assumed that the action radius of the intermolecular forces is negligible, and equals zero.

Let us imagine a volume element V' separated from the body and bounded by surface A'; the interactions between the particles inside and outside the separated volume are transmitted across the surface A'. The transmission of the interactions across the arbitrary element dA located on the surface A', is expressed by the force $\underline{p}\,dA$ and the moment $\underline{m}\,dA$. Consider the point $\underline{x}$ of an elastic body. To determine the stresses acting at this point, let us imagine three coordinate planes passing through this point and perpendicular to the axes of a rectangular Cartesian coordinate system. Let $\underline{p}^{(1)}$ denote the force-stresses vector acting on the surface element $dA_1 = dx_2 dx_3$ and $\underline{m}^{(1)}$ the couple-stress vector. Vectors $\underline{p}^{(1)}$ and $\underline{m}^{(1)}$ and their components, force stresses σ_{1j} and couple stresses μ_{1j} are shown in Fig. 1.2.1. If the stress vectors acting upon the element $dA_2 = dx_1 dx_3$ are denoted by $\underline{p}^{(2)}$, $\underline{m}^{(2)}$, and the vectors acting on the element $dA_3 = dx_1 dx_2$ by $\underline{p}^{(3)}$, $\underline{m}^{(3)}$, the following vector components are obtained:

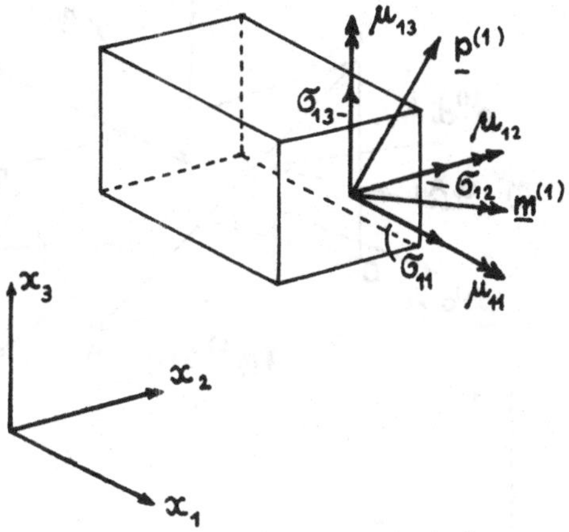

Fig.1.2.1

$$\underline{p}^{(1)} \equiv (\sigma_{11}, \sigma_{12}, \sigma_{13}), \qquad \underline{m}^{(1)} \equiv (\mu_{11}, \mu_{12}, \mu_{13}),$$
$$\underline{p}^{(2)} \equiv (\sigma_{21}, \sigma_{22}, \sigma_{23}), \qquad \underline{m}^{(2)} \equiv (\mu_{21}, \mu_{22}, \mu_{23}), \tag{1.2.1}$$
$$\underline{p}^{(3)} \equiv (\sigma_{31}, \sigma_{32}, \sigma_{33}), \qquad \underline{m}^{(3)} \equiv (\mu_{31}, \mu_{32}, \mu_{33}).$$

Quantities σ_{ji} (i,j = 1,2,3) determine the force-stresses and quantities μ_{ji} the couple stresses.

Consider an infinitesimal tetrahedron OABC acted upon by forces and couples (Fig. 1.2.2) In particular, force-stress σ_{ji} and couple-stress μ_{ji} act on the surface dA, of area ABC. Surface dA_1 of area OBC is loaded by vectors $\underline{p}^{(1)} dA_1$ and $\underline{m}^{(1)} dA_1$. Surface dA_2 is loaded by vectors $\underline{p}^{(2)} dA_2$ and $\underline{m}^{(2)} dA_2$, and surface dA_3, of area OAB, is loaded by vectors $\underline{p}^{(3)}dA_3$ and $\underline{m}^{(3)}dA_3$.

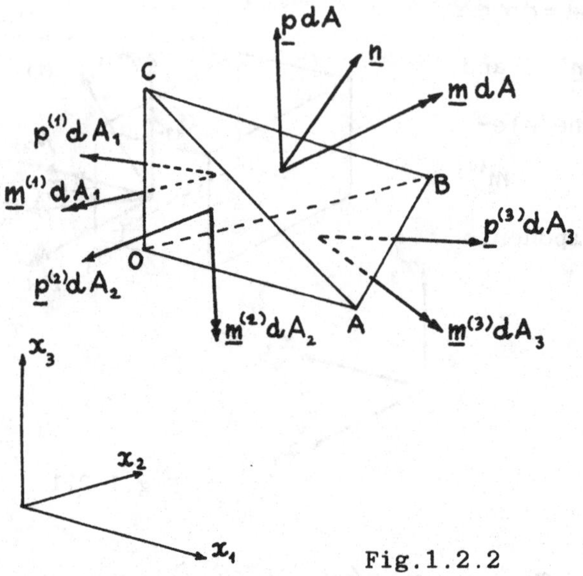

Fig.1.2.2

The element will remain in equilibrium, provided the following equation (1.2.2)

$$
\begin{aligned}
\underline{p}\,dA &= \underline{p}^{(1)}dA_1 + \underline{p}^{(2)}dA_2 + \underline{p}^{(3)}dA_3 \\
\underline{m}\,dA &= \underline{m}^{(1)}dA_1 + \underline{m}^{(2)}dA_2 + \underline{m}^{(3)}dA_3
\end{aligned}
\tag{1.2.2}
$$

The influence of body forces and couples, considered as small, is neglected in the equations of equilibrium. The body forces and couples are proportional to the volume of the tetrahedron. Hence, they are of higher order than the forces acting on the surfaces of the element.

Taking into account that

$$dA_i = dA\,n_i\,, \quad n_i = \cos(\underline{n}, x_i)\,, \quad i = 1, 2, 3 \tag{1.2.3}$$

where $\underline{n}$ is the unit vector normal to the surface element dA, Eq.(1.2.2) takes the form

$$\underline{p} = \underline{p}^{(1)} n_1 + \underline{p}^{(2)} n_2 + \underline{p}^{(3)} n_3$$
$$\underline{m} = \underline{m}^{(1)} n_1 + \underline{m}^{(2)} n_2 + \underline{m}^{(3)} n_3 \, . \tag{1.2.4}$$

The vector equations can be written in the stress components (1.2.1)

$$p_i = \sigma_{ji} n_j \, , \qquad m_i = \mu_{ji} n_j \, . \tag{1.2.5}$$

Let us consider an elastic body remaining in the state of equilibrium after deformation. Separate from the body the region V', bounded by surface A'. The equations of equilibrium of forces and moments for the separated body element have the form

$$\int_{A'} \underline{p} \, dA + \int_{V'} \underline{X} \, dV = 0 \, , \tag{1.2.6}$$

$$\int_{A'} (\underline{r} \times \underline{p} + \underline{m}) \, dA + \int_{V'} (\underline{r} \times \underline{X} + \underline{Y}) \, dV = 0 \, . \tag{1.2.7}$$

Here $\underline{r}$ becomes the position vector with the origin located at a fixed point of the body, e.g. at the origin of the coordinate system. The equations of equilibrium written in a Cartesian rectangular coordinate system have the form

$$\int_{A'} p_i \, dA + \int_{V'} X_i \, dA = 0 \, , \tag{1.2.8}$$

and

$$\int_{A'} (\epsilon_{ijk} x_j p_k + m_i)\, dA + \int_{V'} (\epsilon_{ijk} x_j X_k + Y_i)\, dV = 0 . \tag{1.2.9}$$

Here ϵ_{ijk} is the Levi-Civita alternator: if the permutation of indices is even, ϵ_{ijk} equals +1 and if it is an odd permutation, it is equal to -1. When any two indices are equal, the symbol attains the value zero.

Taking into account relations (1.2.5), Eqs. (1.2.8 - 9) can be written in the form

$$\int_{A'} \sigma_{ji} n_j\, dA + \int_{V'} X_i\, dV = 0 \tag{1.2.10}$$

$$\int_{A'} (\epsilon_{ijk} x_j \sigma_{lk} n_l + \mu_{ji} n_j)\, dA + \int_{V'} (\epsilon_{ijk} X_k x_j + Y_i)\, dV = 0 . \tag{1.2.11}$$

Applying the Ostrogradski-Gauss divergence theorem we are led to

$$\int_{V'} (\sigma_{ji,j} + X_i)\, dV = 0 . \tag{1.2.10'}$$

Due to the arbitrary choice of the volume element V'

$$\sigma_{ji,j} + X_i = 0 . \tag{1.2.12}$$

By means of the Ostrogradski-Gauss divergence theorem and the relation (1.2.12) the Eq.(1.2.11) can be written as

$$\int_{V'} (\epsilon_{ijk} \sigma_{jk} + \mu_{ji,j} + Y_i)\, dV = 0 .$$

Since the volume V' was chosen arbitrarily, one obtains

$$\epsilon_{ijk}\sigma_{jk} + \mu_{ji,j} + Y_i = 0 \ . \qquad (1.2.13)$$

Equations (1.2.12) and (1.2.13) are the equilibrium equations sought for Eq.(1.2.13) indicates that the tensor σ_{ji} is not symmetric, since the asymmetric part of the stress tensor is determined by the body couples and the divergence of the couple-stress tensor. Only in the case when $\mu_{ji} = 0$, $Y_i = 0$, the stress tensor σ_{ji} will be symmetric, since

$$\epsilon_{ijk}\sigma_{jk} = 0, \quad \text{whence} \quad \sigma_{ji} = \sigma_{ij} \ . \qquad (1.2.14)$$

Eq.(1.2.13) can be also written as

$$\epsilon_{ijk}\sigma_{\langle jk\rangle} + \mu_{ji,j} + Y_i = 0 , \qquad (1.2.13')$$

where $\sigma_{\langle jk\rangle}$ is the skew-symmetric part of σ_{jk} .

In the six equilibrium equations (1.2.12 - 13), 18 unknown components of the state of stress appear: 9 components of the tensor σ_{ji} and 9 components of the tensor μ_{ji} .

Let us observe that conditions (1.2.5) written for the points of the surface A bounding the body assume the role of the boundary conditions for the prescribed loadings on A . The equations of motion are derived from the laws of conservation of moment and moment of momentum.

According to Newton's law, the material rate of change of momentum equals the resultant force applied to the body;

$$2.15)\quad \frac{D}{Dt}\int_{V'}\rho v_i\, dV = \int_{A'} p_i\, dA + \int_{V'} X_i\, dV\,, \quad v_i = \dot{v}_i = \frac{\partial u_i}{\partial t}\quad .$$

ere ρ denotes the density of the body, and $\frac{D}{Dt}$ is the material time-derivative which, in the case of infinitesimal deformations of the body considered here, can be replaced by the usual partial derivative $\frac{\partial}{\partial t}$.

Applying the divergence theorem, Eq. (1.2.15) is written in the form

$$\int_{V'}\rho \dot{v}_i\, dV = \int_{V'}\left(\sigma_{ji,j} + X_i\right) dV\quad .$$

Owing to the arbitrary character of the considered volume V' and assuming the integrand to be continuous, the first equation of motion is obtained

$$(1.2.15)\qquad \sigma_{ji,j} + X_i = \rho\, \ddot{u}_i\,, \quad \ddot{u}_i = \frac{\partial^2 u_i}{\partial t^2}\quad .$$

Making use of Euler's equation of motion which states that the material rate of change of momentum H_i with respect to a given point is equal to the resultant moment M_i,

$$(1.2.16)\qquad M_i = \frac{D}{Dt} H_i\ .$$

The ses..cant moment of forces calculated with respect to the or...n .. the reference frame is expressed by the following integral

$$M_i = \int_{A'} (\epsilon_{ijk} x_j p_k + m_i)\, dA + \int_{V'} (\epsilon_{ijk} x_j X_k + Y_i)\, dV .$$

With the aid of the divergence theorem we are led to

$$M_i = \int_{V'} \left[\epsilon_{ijk} x_j (\sigma_{lk,l} + X_k) + (\epsilon_{ijk} \sigma_{jk} + \mu_{ji,j} + Y_i) \right] dV .$$

The body occupying at the instant t the volume V' has the following moment of momentum

$$H_i = \int_{V'} (\rho\, \epsilon_{ijk} x_j v_k + \mathcal{J} \dot{\varphi}_i)\, dV . \qquad (1.2.17)$$

In the above expression the "spin" component $\mathcal{J}\dot{\varphi}_i$ of the moment of momentum, independent of the motion of the point, is added to the usual (orbital) value $\epsilon_{ijk} x_j v_k$ of the moment of momentum. Such a "spin"-like phenomenon can be visualized in a medium; of which the particles are not material points but infinitesimal rigid bodies. The configuration of the body consists of vector triples: each point of the body is connected with three orthonormal direction vectors of the medium. Only in the case of such a medium, with points having six degrees of freedom, the angular velocity of the rotary motion or the spin can be determined. This rotary motion corresponds to the spin moment of momentum.

Returning to the equation of moment of momentum (1.2.16) and assuming the deformations of the body to be small we are led to

$$\int_{V'}\left[\epsilon_{ijk}x_j\left(\sigma_{lk,l}+X_k-\rho\ddot{u}_k\right)+\left(\epsilon_{ijk}\sigma_{jk}+\mu_{ji,j}+Y_i-\mathcal{J}\ddot{\varphi}_i\right)\right]dV=0\,.$$

Taking into account the equations of motion (1.2.15), the above formula can be reduced to the local equation

$$\epsilon_{ijk}\sigma_{jk}+\mu_{ji,j}+Y_i=\mathcal{J}\ddot{\varphi}_i\,. \qquad (1.2.18)$$

Equations of motion (1.2.15) and (1.2.18) are to be supplemented by the boundary and initial conditions. The latter equations have the form

$$\begin{aligned}&u_i(\underline{x},0)=f_i(\underline{x}),\quad \dot{u}_i(\underline{x},0)=g_i(\underline{x}),\\ &\varphi_i(\underline{x},0)=h_i(\underline{x}),\quad \dot{\varphi}_i(\underline{x},0)=k(\underline{x}),\quad \underline{x}\in V\,.\end{aligned} \qquad (1.2.19)$$

This means that at the instant $t=0$ the displacements, the rates of displacements, the rotations and their time-rates of change have to be prescribed.

The aim of our considerations is to express the equations of motion in terms of the components of the displacements and rotation vectors. This leads to a system of six equations involving the six unknown functions - the components of the displacement vector $\underline{u}$ and the rotation vector $\underline{\varphi}$.

1. 3 Conservation of Energy, Entropy Balance.

Let the body undergo certain deformations under the influence of the time-dependent external loads. It is assum ed that heat sources exist within the body and that heat is absorbed from the surroundings.

During the process of deformation heat is exchanged between the elements of the body owing to the heat conduction. Let us apply the law of conservation of energy to the volume V bounded by the surface A.

$$\frac{d}{dt}(\mathcal{U}+\mathcal{K}) = \mathcal{L} + \dot{Q} . \tag{1.3.1}$$

Here $\mathcal{U}$ is the internal energy, $\mathcal{K}$ - kinetic energy, $\mathcal{L}$ power of external forces, and $\dot{Q}$ - non-mechanical power, the rate of heat input due to the heat flux and heat production by the source.

The mechanical power is expressed by the formula

$$\mathcal{L} = \int_A (p_i v_i + m_i w_i)\, dA + \int_V (X_i v_i + Y_i w_i)\, dV, \tag{1.3.2}$$

where

$$v_i = \dot{u}_i , \qquad w_i = \dot{\varphi}_i .$$

The kinetic energy $\mathcal{K}$ is described by the integ-

+)The considerations presented here are more extensively treated (in connection with the classical elastic continuum) by this author in the monograph "Dynamical Problems of Thermoelasticity" (in Polish) Warsaw, PWN. 1968.

ral

$$\mathcal{K} = \frac{1}{2}\int_V (\rho v_i v_i + \mathcal{J} w_i w_i)\, dV . \tag{1.3.3}$$

The conservation law (1.3.1) is written in the form

$$\frac{d}{dt}\int \left[U + \frac{1}{2}(\rho v_i v_i + \mathcal{J} w_i w_i)\right] dV =$$

$$= \int_V (X_i v_i + Y_i w_i)\, dV + \int_A (p_i v_i + m_i w_i)\, dA - \int_A q_i n_i \, dA . \tag{1.3.4}$$

Here q_i is the heat flux component and U denotes the internal energy per unit volume. The left-hand side expression of Eq. (1.3.4) represents the time-rate of change of the kinetic and internal energy. The first right-hand side term represents the power of body forces and couples, the second term - the power of the surface forces and couples. Finally, the last term corresponds to the amount of heat transported into the volume V by heat conduction. Applying the divergence theorem to Eq. (1.3.4) and taking into account the equations of motion

$$\sigma_{ji,j} + X_i = \rho \ddot{v}_i , \quad \epsilon_{ijk}\sigma_{jk} + \mu_{ji,j} + Y_i = \mathcal{J}\ddot{\varphi}_i , \tag{1.3.5}$$

the following equation is obtained

$$\int_V \left\{ \dot{U} - \left[\sigma_{ji}(v_{i,j} - \epsilon_{kji} w_k) + \mu_{ji} w_{i,j}\right] + q_{i,i} \right\} dV = 0 . \tag{1.3.6}$$

This equation is satisfied for arbitrary volume V . If the integrand is continuous, relation

$$\dot{U} = \sigma_{ji}\dot{\gamma}_{ji} + \mu_{ji}\dot{\varkappa}_{ji} - q_{i,i} , \tag{1.3.7}$$

with

$$\gamma_{ji} = u_{i,j} - \epsilon_{kji}\varphi_k \ , \qquad \varkappa_{ji} = \varphi_{i,j} \ , \tag{1.3.8}$$

is locally satisfied. γ_{ji} denotes the asymmetric strain tensor, and $\varkappa_{ji}$ is the torsion tensor.

The equation of the entropy balance has the form[+)]

$$\int_V \dot{S}\, dV = -\int_A \frac{q_i n_i}{T}\, dA + \int_V \Theta\, dV \ . \tag{1.3.9}$$

Here S is the specific entropy (per unit volume). The left-hand side represents the change of entropy: the first right-hand side term represents the increase of entropy, resulting from its interchange with the surroundings, the second term - the entropy production, resulting from the heat transfer.

Applying the divergence theorem we are led to

$$\int_V \left[\dot{S} - \Theta - \left(\frac{q_i}{T} \right)_{,i} \right] dV = 0 \ . \tag{1.3.10}$$

For an arbitrary volume V, and continuous integrand, the following condition holds locally

$$\dot{S} = \Theta - \frac{q_{i,i}}{T} + \frac{q_i T_{,i}}{T^2} \ . \tag{1.3.11}$$

According to the postulate of irreversible thermodynamics, inequality $\Theta \geqslant 0$ should be fulfilled.

+) W. Nowacki, loc.cit., page 5

Elimination of $q_{i,i}$ from Eqs.(1.3.7),(1.3.10) and introduction of the Helmholtz free energy $F = U - ST$ makes it possible to write

$$\dot{F} = \sigma_{ji}\dot{\gamma}_{ji} + \mu_{ji}\dot{\varkappa}_{ji} - \dot{T}S - T\left(\Theta + \frac{q_i T_{,i}}{T^2}\right). \tag{1.3.12}$$

The free energy is a function of independent variables $\gamma_{ji}, \varkappa_{ji}, T$; hence,

$$\dot{F} = \frac{\partial F}{\partial \gamma_{ji}}\dot{\gamma}_{ji} + \frac{\partial F}{\partial \varkappa_{ji}}\dot{\varkappa}_{ji} + \frac{\partial F}{\partial T}\dot{T}. \tag{1.3.13}$$

Let us assume that functions Θ, q_i ,...., σ_{ji}, μ_{ji} do not explicitly depend on the time derivatives of functions γ_{ji}, $\varkappa_{ji}$, T. and define the entropy as $S = -\frac{\partial F}{\partial T}$; then, comparing Eqs. (1.3.12) and (1.3.13) the following relations are obtained

$$\sigma_{ji} = \frac{\partial F}{\partial \gamma_{ji}}, \quad \mu_{ji} = \frac{\partial F}{\partial \varkappa_{ji}},$$
$$S = -\frac{\partial F}{\partial T}, \quad \Theta = -\frac{q_i T_{,i}}{T^2}. \tag{1.3.14}$$

The second law of thermodynamics is satisfied when $\Theta \geqslant 0$, and it follows that

$$-\frac{T_{,i}q_i}{T^2} \geqslant 0.$$

This inequality is satisfied by the Fourier heat-conduction law

$$-q_i = k_{ij}T_{,j}, \quad -q_i = k_{ij}\theta_{,j}, \quad T = T_0 + \theta. \tag{1.3.15}$$

Here the notions of the natural state temperature T_0 and the temperature variation ϑ , have been introduced.

Eqs.(1.3.11) aided by the fourth relation of (1.3.14) and (1.3.15) lead to

$$T\dot{S} = -q_{i,i} = k_{ij}\vartheta_{,ij} . \tag{1.3.16}$$

The entropy source Θ is a quadratic, positive definite function; hence, the quantities k_{ij} should satisfy the following inequalities:

$$k_{jj} \geqslant 0 \ , \quad k_{jj}k_{kk} - k_{jk}k_{kj} \geqslant 0 , \qquad j,k = 1,2,3 .$$

For an isotropic body, Eq.(1.3.16) takes the form

$$T\dot{S} = k\vartheta_{,jj} \ , \tag{1.3.17}$$

where k is the heat conductivity coefficient and is a constant.

1. 4 Constitutive Equations.

Let us expand the free energy function $F(\gamma_{ji}, \varkappa_{ji}, T)$ in the neighbourhood of the natural state: $\gamma_{ji} = 0, \varkappa_{ji} = 0, T = T_0$, into a Taylor series, neglecting the terms of order larger than two in the expressions for atrains.

In an isotropic, homogeneous and centrosymmetric (invariant with respect to rotations) body the following expansion is obtained

$$F = \frac{\mu+\alpha}{2}\gamma_{ji}\gamma_{ji} + \frac{\mu-\alpha}{2}\gamma_{ji}\gamma_{ij} + \frac{\lambda}{2}\gamma_{kk}\gamma_{nn} +$$

$$\frac{\gamma+\varepsilon}{2}\varkappa_{ji}\varkappa_{ji} + \frac{\gamma-\varepsilon}{2}\varkappa_{ji}\varkappa_{ij} + \frac{\beta}{2}\varkappa_{kk}\varkappa_{nn} +$$

$$-\nu\gamma_{kk}\Theta - \chi\varkappa_{kk}\Theta + G(\Theta). \tag{1.4.1}$$

The form of free energy presented here is justified in the following manner: the free energy is a scalar function and each right-hand term should be also a scalar; now, three independent second order invariants can be constructed from the components of the tensor γ_{ji}, namely $\gamma_{ji}\gamma_{ji}$, $\gamma_{ji}\gamma_{ij}$ and $\gamma_{kk}\gamma_{nn}$. The same is true with respect to the tensor $\varkappa_{ji}$. Terms $\gamma_{ji}\varkappa_{ji}$, $\gamma_{ji}\varkappa_{ij}$ and $\gamma_{kk}\varkappa_{nn}$, do not appear in Eq.(1.4.1) since this would contradict the postulate of centro-symmetry. Invariants γ_{kk} and $\varkappa_{kk}$, appear in the seventh and eighth term of expression (1.4.1) which derives from the fact that one invariant of the first kind can be formed from each of the tensors γ_{ji} and $\varkappa_{ji}$, namely γ_{kk} and $\varkappa_{kk}$. Upon using the relations

$$\sigma_{ji} = \frac{\partial F}{\partial \gamma_{ji}}, \quad \mu_{ji} = \frac{\partial F}{\partial \varkappa_{ji}}, \quad S = -\frac{\partial F}{\partial T}. \tag{1.4.2}$$

the following constitutive equations are derived:

$$\begin{aligned} \sigma_{ji} &= (\mu+\alpha)\gamma_{ji} + (\mu-\alpha)\gamma_{ij} + (\lambda\gamma_{kk} - \nu\Theta)\delta_{ji}, \\ \mu_{ji} &= (\gamma+\varepsilon)\varkappa_{ji} + (\gamma-\varepsilon)\varkappa_{ij} + (\beta\varkappa_{kk} - \chi\Theta)\delta_{ji}, \\ S &= \nu\gamma_{kk} + \chi\varkappa_{kk} - \frac{\partial G(\Theta)}{\partial T}. \end{aligned} \tag{1.4.3}$$

The first two relations of (1.4.3) can be rewritten in the equivalent form

$$\sigma_{ij} = 2\mu\gamma_{(ij)} + 2\alpha\gamma_{\langle ij\rangle} + (\lambda\gamma_{kk} - \nu\Theta)\delta_{ij},$$
$$\mu_{ij} = 2\gamma\varkappa_{(ij)} + 2\varepsilon\varkappa_{\langle ij\rangle} + (\beta\varkappa_{kk} - \chi\Theta)\delta_{ij}. \tag{1.4.4}$$

Here μ, λ are the Lamé constants and $\alpha, \beta, \gamma, \varepsilon$ are the new elastic constants. These quantities are referred to the isothermal state. Constants ν , χ depend upon both the mechanical and thermal properties of the body. Symbols ( ) and < > denote the symmetric and skew-symmetric (antisymmetric) part of the tensor.

Equations (1.4.4) can be solved for γ_{ji} and $\varkappa_{ji}$ to give

$$\gamma_{ij} = \alpha_t\Theta\delta_{ij} + 2\mu'\sigma_{(ij)} + 2\alpha'\sigma_{\langle ij\rangle} + \lambda'\delta_{ij}\sigma_{kk},$$
$$\varkappa_{ij} = \beta_t\Theta\delta_{ij} + 2\gamma'\mu_{(ij)} + 2\varepsilon'\mu_{\langle ij\rangle} + \beta'\delta_{ij}\mu_{kk}, \tag{1.4.5}$$

where

$$2\mu' = \frac{1}{2\mu}, \quad 2\alpha' = \frac{1}{2\alpha}, \quad 2\gamma' = \frac{1}{2\gamma}, \quad 2\varepsilon' = \frac{1}{2\varepsilon}, \quad \lambda' = -\frac{\lambda}{6\mu K},$$
$$\beta' = -\frac{\beta}{6\gamma\Omega}, \quad \alpha_t = \frac{\nu}{3K}, \quad \beta_t = \frac{\chi}{3\Omega}, \quad K = \lambda + \frac{2}{3}\mu, \quad \Omega = \beta + \frac{2}{3}\gamma.$$

Consider an infinitesimal element of the body free of stresses σ_{ji} and μ_{ji} on its surface. Then

$$\gamma^{\circ}_{ij} = \alpha_t\Theta\delta_{ij}, \qquad \varkappa^{\circ}_{ij} = \beta_t\Theta\delta_{ij}. \tag{1.4.6}$$

It is known, however, that the temperature change produces rotation-free deformation of the element, thus $\beta_t = 0$ and $\chi = 0$.

α_t is the coefficient of linear thermal expansion.

Now we are able to write down the final form of the constitutive equations*)

$$\sigma_{ij} = 2\mu\gamma_{(ij)} + 2\alpha\gamma_{\langle ij\rangle} + (\lambda\gamma_{kk} - \nu\theta)\delta_{ij},$$

$$\mu_{ij} = 2\gamma\varkappa_{(ij)} + 2\varepsilon\varkappa_{\langle ij\rangle} + \beta\varkappa_{kk}\delta_{ij}, \tag{1.4.7}$$

$$S = \nu\gamma_{kk} - \frac{\partial G(\theta)}{\partial T}.$$

Owing to the fact that $S = S(\gamma_{ji}, \varkappa_{ji}, T)$, we can write

$$dS = \left(\frac{\partial S}{\partial \gamma_{ij}}\right)_{\varkappa,T} d\gamma_{ji} + \left(\frac{\partial S}{\partial \varkappa_{ji}}\right)_{\gamma,T} d\varkappa_{ji} + \left(\frac{\partial S}{\partial T}\right)_{\gamma,\varkappa} dT.$$

The quantity $T = \left(\frac{\partial S}{\partial T}\right)$ measures the amount of heat produced in a unit volume of the body in the course of the temperature change, the deformation of the body being fixed. This quantity is denoted by c_ε and is called the specific heat at constant deformation.

On the other hand, from Eq. (1.4.7) it follows that

$$dS = \nu\, d\gamma_{kk} - \frac{\partial^2 G(\theta)}{\partial T^2}\, dT,$$

---

*)W. Nowacki: "Couple-stresses in the theory of thermoelasticity", Proc. of the IUTAM-Symposium on Irreversible Aspects in Continuum Mechanics, Vienna,1966, Springer Verlag,Wien-New York,1968

and hence,

$$c_\varepsilon = T\left(\frac{\partial S}{\partial T}\right)_{\gamma,\varkappa} = -T\,\frac{\partial^2 G(\vartheta)}{\partial T^2}\ .$$

Let us integrate this equation twince; the integration constants vanish since for a natural state $S=0$, $F=0$. Then we have

$$G(\vartheta) = -\int_{T_0}^{T} dT \int_{T_0}^{T} \frac{c_\varepsilon\, dT}{T}\ .$$

whence,

$$-\frac{\partial G(\vartheta)}{\partial T} = c_\varepsilon \ln \frac{T}{T_0}\ . \tag{1.4.10}$$

In such a way the third expression for the entropy (1.4.7) can be written in the form

$$S = \nu\gamma_{kk} + c_\varepsilon \ln\left(1 + \frac{\vartheta}{T_0}\right). \tag{1.4.11}$$

Making the assumption that $\left|\frac{\vartheta}{T_0}\right| \ll 1$ - what confines our considerations to small temperature intervals, function $\ln\left(1 + \frac{\vartheta}{T_0}\right)$ can be expanded into a series in which only the first term is preserved

$$S = \nu\gamma_{kk} + \frac{c_\varepsilon}{T}\,\vartheta\ . \tag{1.4.12}$$

1.5 Fundamental Relations and Equations of Elastokinetics.

The considerations presented in sections 1.3 and 1.4 concern the so-called "coupled thermoelasticity". In the following sections of this chapter the simplified theory of time-dependent deformations of the body will be applied - the Cosserat continuum elastokinetics.

Let us make the assumption that the heat sources are absent from the body and that the body does not absorb the heat from the surroundings. Assume, moreover, that the exchange (caused by conduction) proceeds very slowly during the process of deformation of the body. If the heat exchange is practically negligible in time intervals of the order of vibration periods, each part of the body can be considered as thermally insulated, and the entire process - as a thermodynamic adiabatic process. In this case the assumption $\underline{q} = 0$, $(Q = 0)$ should be made in the energy conservation law. Hence we are led to

$$\frac{d}{dt}\left(\mathcal{U} + \mathcal{K}\right) = \mathcal{L}, \tag{1.5.1}$$

whence, the local equation can be derived

$$\dot{U} = \sigma_{ji}\dot{\gamma}_{ji} + \mu_{ji}\dot{\varkappa}_{ji}\,. \tag{1.5.2}$$

The equation of the entropy balance, with $q_i = 0$, is identically satisfied, provided $\dot{S} = 0$.

Owing to the relations

$$S = \nu\gamma_{kk} + \frac{c_\varepsilon}{T_0}\vartheta, \quad \dot{S} = \nu\dot{\gamma}_{kk} + \frac{c_\varepsilon}{T_0}\dot{\vartheta} = 0,$$

it follows that

$$\vartheta = -\frac{\nu T_0}{c_\varepsilon}\gamma_{kk} = -\frac{\nu T_0}{c_\varepsilon}\operatorname{div}\underline{u}\,. \tag{1.5.3}$$

In the course of an adiabatic process, temperature $\vartheta = T - T_0$ is proportional to the dilatation $\operatorname{div}\underline{u}$. These considerations can be generalized to anisotropic and not centro-symmetric bodies. The internal energy function $U(\gamma_{ji}, \varkappa_{ji})$ is expanded into the Mac Laurin series in the neighbourhood of the natural state

$$U(\gamma_{ji}, \varkappa_{ji}) = U_0 + \left(\frac{\partial U}{\partial \gamma_{ji}}\right)_0 \gamma_{ji} + \left(\frac{\partial U}{\partial \varkappa_{ji}}\right)_0 \varkappa_{ji} + \frac{1}{2}\left[\left(\frac{\partial^2 U}{\partial \gamma_{ji}\partial \gamma_{\ell k}}\right)_0 \gamma_{ji}\gamma_{\ell k} + \right.$$

$$\left. + 2\left(\frac{\partial^2 U}{\partial \gamma_{ji}\partial \varkappa_{k\ell}}\right)_0 \gamma_{ji}\varkappa_{k\ell} + \left(\frac{\partial^2 U}{\partial \varkappa_{ji}\partial \varkappa_{k\ell}}\right)_0 \varkappa_{ji}\varkappa_{k\ell}\right] + \ldots.. \tag{1.5.4}$$

In order to confine the stress-strain laws to linear relations, only the quadratic terms in Eq.(1.5.4) are retained. Relations

$$\frac{\partial U}{\partial \gamma_{ji}} = \sigma_{ji}\,, \qquad \frac{\partial U}{\partial \varkappa_{ji}} = \mu_{ji}\,, \tag{1.5.5}$$

yield

$$\sigma_{ji} = a_{jik\ell}\,\gamma_{k\ell} + b_{jik\ell}\,\varkappa_{k\ell}\,, \tag{1.5.6}$$

$$\mu_{ji} = b_{k\ell ji}\,\gamma_{k\ell} + c_{jik\ell}\,\varkappa_{k\ell}\,. \tag{1.5.7}$$

New notations have been introduced here

$$\left(\frac{\partial^2 U}{\partial \gamma_{ji}\partial \gamma_{k\ell}}\right)_0 = a_{jik\ell}\,, \qquad \left(\frac{\partial^2 U}{\partial \gamma_{ji}\partial \varkappa_{k\ell}}\right)_0 = b_{jik\ell}\,, \qquad \left(\frac{\partial^2 U}{\partial \varkappa_{ji}\partial \varkappa_{k\ell}}\right)_0 = c_{jik\ell}\,.$$

Since, the natural state, the strains $\gamma_{ji} = 0$, $\varkappa_{ji} = 0$, have to imply the state of stress $\sigma_{ji} = 0$, $\mu_{ji} = 0$, the coefficients multiplying the linear terms of the state of strain should vanish in the general expression for the internal energy. The generality of our considerations will not be affected if the assumption $U_0 = 0$ is made.

It follows that internal energy has the form

$$U = \frac{1}{2} a_{jikl} \gamma_{ji} \gamma_{kl} + \frac{1}{2} c_{jikl} \varkappa_{ji} \varkappa_{kl} + b_{jikl} \gamma_{ji} \varkappa_{kl} . \tag{1.5.7}$$

Relations (1.5.6),(1.5.7) represent now the constitutive equations for an anisotropic micropolar medium, and formula (1.5.7) the internal energy per unit volume of the body.

From the fact that dU is a total differential we can drow the conclusion that

$$\frac{\partial^2 U}{\partial \gamma_{ji} \partial \gamma_{kl}} = \frac{\partial^2 U}{\partial \gamma_{kl} \partial \gamma_{ji}} \quad \text{or} \quad \frac{\partial \sigma_{ji}}{\partial \gamma_{kl}} = \frac{\partial \sigma_{kl}}{\partial \gamma_{ji}} ,$$

(1.5.8)

$$\frac{\partial^2 U}{\partial \varkappa_{ji} \partial \varkappa_{kl}} = \frac{\partial^2 U}{\partial \varkappa_{kl} \partial \varkappa_{ji}} \quad \text{or} \quad \frac{\partial \mu_{ji}}{\partial \varkappa_{kl}} = \frac{\partial \mu_{kl}}{\partial \varkappa_{ji}} .$$

The derived thermodynamic relations yield the conditions of symmetry of tensors a_{jikl} and c_{jikl} :

$$a_{jikl} = a_{klji} , \qquad c_{jikl} = c_{klji} .$$

Tensor b_{jikl} does not possess this property; it forms a pseudo-tensor, inversion of the coordinate system changing its sign. In a centrosymmetric body all components of b_{jikl} vanish.

In what follows, only centrosymmetric medium is dealt with; hence

$$U = \frac{1}{2} a_{jikl} \gamma_{ji} \gamma_{kl} + \frac{1}{2} c_{jikl} \varkappa_{ji} \varkappa_{kl} , \tag{1.5.9}$$

and

$$\sigma_{ji} = a_{jikl} \gamma_{kl} , \quad \mu_{ji} = c_{jikl} \varkappa_{kl} . \tag{1.5.9'}$$

Passing to isotropic bodies, the coefficients a_{jikl} , c_{jikl} are required to remain invariant to the changes of the coordinate system. This requirement is fulfilled if the quantities are written in the form of a general isotropic tensor of rank four,

$$\begin{aligned} a_{jikl} &= (\mu + \alpha) \delta_{jk} \delta_{il} + (\mu - \alpha) \delta_{jl} \delta_{ik} + \lambda \delta_{ij} \delta_{kl} , \\ c_{jikl} &= (\gamma + \varepsilon) \delta_{jk} \delta_{il} + (\gamma - \varepsilon) \delta_{jl} \delta_{ik} + \beta \delta_{ij} \delta_{kl} . \end{aligned} \tag{1.5.10}$$

Here μ , λ , α , β , γ , ε are material constants (for adiabatic processes in isotropic bodies).

Insertion of Eq.(1.5.10) in (1.5.9) leads to the expression for the internal energy

$$\begin{aligned} U = {} & \frac{\mu + \alpha}{2} \gamma_{ji} \gamma_{ji} + \frac{\mu - \alpha}{2} \gamma_{ji} \gamma_{ij} + \frac{\lambda}{2} \gamma_{kk} \gamma_{nn} + \\ & + \frac{\gamma + \varepsilon}{2} \varkappa_{ji} \varkappa_{ji} + \frac{\gamma - \varepsilon}{2} \varkappa_{ji} \varkappa_{ij} + \frac{\beta}{2} \varkappa_{kk} \varkappa_{nn} . \end{aligned} \tag{1.5.11}$$

Equations (1.5.9) are written in the form

(1.5.12) $$\sigma_{ji} = (\mu + \alpha)\gamma_{ji} + (\mu - \alpha)\gamma_{ij} + \lambda\delta_{ji}\gamma_{kk},$$

(1.5.13) $$\mu_{ji} = (\gamma + \varepsilon)\varkappa_{ji} + (\gamma - \varepsilon)\varkappa_{ij} + \beta\delta_{ji}\varkappa_{kk}.$$

Material constants μ, λ, α, β, γ, ε appearing in these formulas concern the adiabatic processes measured under adiabatic conditions. Relations (1.5.12),(1.5.13) can be derived from the general formula (1.3.3):

(1.5.12') $$\sigma_{ji} = (\mu_T + \alpha_T)\gamma_{ji} + (\mu_T - \alpha_T)\gamma_{ij} + (\lambda_T\gamma_{kk} - \nu_T\theta)\delta_{ji},$$

(1.5.13') $$\mu_{ji} = (\gamma_T + \varepsilon_T)\varkappa_{ji} + (\gamma_T - \varepsilon_T)\varkappa_{ij} + \beta_T\varkappa_{kk}\delta_{ji}$$

Here μ_T, λ_T, α_T, β_T, γ_T, ε_T denote material constants measured during an isothermal process. By inserting (1.5.3) in (1.5.12') and (1.5.13'), we are led to relations (1.5.12) and (1.5.13), provided the substitutions

$$\mu = \mu_T, \qquad \alpha = \alpha_T, \qquad \beta = \beta_T, \qquad \gamma = \gamma_T, \qquad \varepsilon = \varepsilon_T$$

and

$$\lambda = \lambda_T + \nu_T^2\frac{T_0}{T},$$

are made. The values of the constants μ, α, β, γ, ε in both the adiabatic and isothermal processes are evidently the same. The only difference appears in the values of λ and λ_T.

Let us reconsider the expression for the internal energy (1.5.11). By splitting tensors γ_{ji} , $\varkappa_{ji}$ into symmetric and antisymmetric parts, we obtain

$$U = \mu\,\gamma_{(ij)}\gamma_{(ij)} + \alpha\,\gamma_{\langle ij\rangle}\gamma_{\langle ij\rangle} + \frac{\lambda}{2}\gamma_{kk}\gamma_{nn} + \\ + \gamma\,\varkappa_{(ij)}\varkappa_{(ij)} + \varepsilon\,\varkappa_{\langle ij\rangle}\varkappa_{\langle ij\rangle} + \frac{\beta}{2}\varkappa_{kk}\varkappa_{nn}\,, \qquad (1.5.14)$$

and

$$\sigma_{ij} = 2\mu\,\gamma_{(ij)} + 2\alpha\,\gamma_{\langle ij\rangle} + \lambda\,\gamma_{kk}\delta_{ij}\,, \qquad (1.5.15)$$

$$\mu_{ij} = 2\gamma\,\varkappa_{(ij)} + 2\varepsilon\,\varkappa_{\langle ij\rangle} + \beta\,\varkappa_{kk}\delta_{ij}\,. \qquad (1.5.16)$$

By decomposing σ_{ij} and μ_{ij} into symmetric and antisymmetric parts, we obtain

$$\sigma_{(ij)} = 2\mu\,\gamma_{(ij)} + \lambda\,\gamma_{kk}\delta_{ij}\,, \qquad \sigma_{\langle ij\rangle} = 2\alpha\,\gamma_{\langle ij\rangle} \qquad (1.5.17)$$

$$\mu_{(ij)} = 2\gamma\,\varkappa_{(ij)} + \beta\,\varkappa_{kk}\delta_{ij}\,, \qquad \mu_{\langle ij\rangle} = 2\varepsilon\,\varkappa_{\langle ij\rangle}\,. \qquad (1.5.18)$$

The first form of Eq.(1.5.17) coincides exactly with Hooke's law of the classical theory.

Contraction of tensors σ_{ji} and μ_{ji} yields

$$\sigma_{kk} = 3K\gamma_{kk}\,, \qquad \mu_{kk} = 3L\,\varkappa_{kk}\,, \qquad (1.5.19)$$

where

$$K = \lambda + \frac{2}{3}\mu\,, \qquad L = \beta + \frac{2}{3}\gamma\,.$$

Equations (1.5.12),(1.5.13) can be solved with respect to $\varkappa_{ji}$, γ_{ji}. Then

(1.5.20) $$\gamma_{ij} = 2\mu' \sigma_{(ij)} + \lambda' \delta_{ij} \sigma_{kk} + 2\alpha' \sigma_{\langle ij \rangle} ,$$

(1.5.21) $$\varkappa_{ij} = 2\gamma' \mu_{(ij)} + \beta' \delta_{ij} \mu_{kk} + 2\varepsilon' \mu_{\langle ij \rangle} ,$$

where

$$2\mu' = \frac{1}{2\mu} , \quad 2\gamma' = \frac{1}{2\gamma} , \quad 2\alpha' = \frac{1}{2\alpha} , \quad 2\varepsilon' = \frac{1}{2\varepsilon} ,$$

$$\lambda' = -\frac{\lambda}{6\mu K} , \quad \beta' = -\frac{\beta}{6\gamma L} .$$

Assume the uniform hydrostatic pressure:

$$\sigma_{ij} = -p\,\delta_{ij} , \qquad p > 0 .$$

From Eq.(1.5.19) it follows that

$$\gamma_{kk} = -\frac{p}{K} .$$

The volume of the body decreases; the coefficient $K > 0$ can be recognized as the well-known classical elasticity bulk modulus.

Under the assumption that the body is subjected to uniform torsion $\mu_{ij} = -m\,\delta_{ij}$, $m > 0$, Eq.(1.5.19) leads to

$$\varkappa_{kk} = -\frac{m}{L} .$$

The quantity L can be treated as the measure

of twist of the body.

Material constants μ, λ, α, β, γ, ε should satisfy certain limitations coming from the fact that the internal energy function is a positive definite quadratic function. The internal energy can be represented in a slightly different form: by numbering properly the quantities γ_{ji} and $\varkappa_{ji}$ and denoting them by y_α ($\alpha = 1,2,\ldots,17,18$), so that

$$2U = a_{\alpha\beta}\, y_\alpha y_\beta$$

and assuming here the following notations

$$\gamma_{11} = y_1, \quad \gamma_{22} = y_2, \quad \gamma_{33} = y_3, \quad \varkappa_{11} = y_4, \quad \varkappa_{22} = y_5, \quad \varkappa_{33} = y_6,$$

$$\gamma_{31} = y_7, \quad \gamma_{13} = y_8, \quad \varkappa_{13} = y_9, \quad \varkappa_{31} = y_{10}, \quad \gamma_{21} = y_{11}, \quad \varkappa_{12} = y_{12},$$

$$\gamma_{12} = y_{13}, \quad \varkappa_{21} = y_{14}, \quad \gamma_{23} = y_{15}, \quad \varkappa_{23} = y_{16}, \quad \gamma_{32} = y_{17}, \quad \varkappa_{32} = y_{18},$$

the matrix of coefficients can be written in the form

$$\| a_{\alpha\beta} \| = \begin{Vmatrix} \mathrm{I} & & & & \\ & \mathrm{II} & & & \\ & & \mathrm{III} & & \\ & & & \mathrm{IV} & \\ & & & & \mathrm{V} \end{Vmatrix} .$$

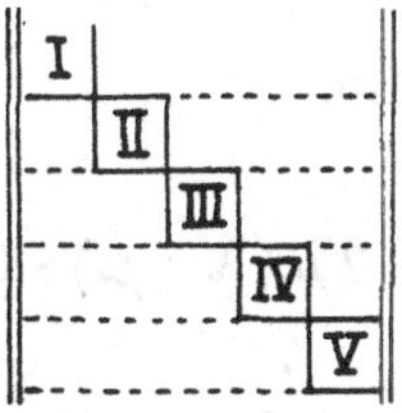

The determinant Δ of this matrix can be written in the form of a product of five determinants

$$\Delta = \Delta_{\mathrm{I}} \cdot \Delta_{\mathrm{II}} \cdot \Delta_{\mathrm{III}} \cdot \Delta_{\mathrm{IV}} \cdot \Delta_{\mathrm{V}}$$

where

$$\Delta_{I} = \begin{vmatrix} \lambda+2\mu & \lambda & \lambda \\ \lambda & \lambda+2\mu & \lambda \\ \lambda & \lambda & \lambda+2\mu \end{vmatrix} \qquad \Delta_{II} = \begin{vmatrix} \beta+2\gamma & \beta & \beta \\ \beta & \beta+2\gamma & \beta \\ \beta & \beta & \beta+2\gamma \end{vmatrix}$$

$$\Delta_{III} = \begin{vmatrix} \mu-\alpha & \mu+\alpha & 0 & 0 \\ \mu-\alpha & \mu+\alpha & 0 & 0 \\ 0 & 0 & \gamma+\varepsilon & \gamma-\varepsilon \\ 0 & 0 & \gamma-\varepsilon & \gamma-\varepsilon \end{vmatrix} \qquad \Delta_{IV} = \Delta_{V} = \begin{vmatrix} \mu+\alpha & 0 & \mu-\alpha & 0 \\ 0 & \gamma+\varepsilon & 0 & \gamma-\varepsilon \\ \mu-\alpha & 0 & \mu+\alpha & 0 \\ 0 & \gamma-\varepsilon & 0 & \gamma+\varepsilon \end{vmatrix} .$$

According to the Sylvester theorem, the necessary and sufficient condition for a quadratic form to be positive definite is the fulfillment of the two inequalities

$$M_1 = a_{11} > 0 , \quad M_\ell = \begin{vmatrix} a_{11} & a_{12} \cdots\cdots a_{1\ell} \\ a_{21} & a_{22} \cdots\cdots a_{2\ell} \\ \cdots\cdots & \cdots\cdots \\ a_{\ell 1} & a_{\ell 2} \cdots\cdots a_{\ell\ell} \end{vmatrix} > 0 , \quad \ell = 2, 3, \ldots, n .$$

Determinant Δ_I yields

$$M_1^{(I)} = \lambda + 2\mu > 0 , \quad M_2^{(I)} = 2\mu(\lambda + \mu) > 0 , \quad M_3^{(I)} = 4\mu^2(3\lambda + 2\mu) > 0 .$$

Two conditions for the coefficients are obtained

$$\mu > 0 , \qquad 3\lambda + 2\mu > 0 .$$

Determinant Δ_{II} yields in an analogous way

$$\gamma > 0 , \qquad 3\beta + 2\gamma > 0 .$$

Further inequalities follow from the application of the Sylvester theorem to the determinant Δ_{III} The corresponding inequalities read

$$\mu + \alpha > 0\,, \quad \gamma + \varepsilon > 0\,, \quad \alpha > 0\,, \quad \varepsilon > 0\,.$$

No additional conditions follow from the remaining determinants

$$\Delta_{IV}\,, \qquad \Delta_{V}\,.$$

Let us reconsider the expression for the internal energy U Eq.(1.5.14) and the stress-strain relations (1.5.15 - 16) and assume $\alpha = 0$. In a body characterized by five material constants

$$U = \mu \gamma_{(ij)} \gamma_{(ij)} + \frac{\lambda}{2} \gamma_{kk} \gamma_{nn} + \gamma \varkappa_{(ij)} \varkappa_{(ij)} + \varepsilon \varkappa_{\langle ij \rangle} \varkappa_{\langle ij \rangle} + \frac{\beta}{2} \varkappa_{kk} \varkappa_{nn} \qquad (1.5.22)$$

and

$$\sigma_{ij} = 2\mu \gamma_{(ij)} + \lambda \gamma_{kk} \delta_{ij} \qquad (1.5.23)$$

$$\mu_{ij} = 2\gamma \varkappa_{(ij)} + 2\varepsilon \varkappa_{\langle ij \rangle} + \beta \varkappa_{kk} \delta_{ij}\,, \quad \gamma_{(ij)} = \frac{1}{2}\left(u_{i,j} + u_{j,i}\right). \qquad (1.5.24)$$

We are led to a symmetric stress tensor σ_{ij} and an asymmetric couple stress tensor μ_{ij} . The passage to the classical theory of elasticity is possible if the material constants γ , ε , β are assumed to vanish; then we obtain

$$U = \mu \gamma_{(ij)} \gamma_{(ij)} + \frac{\lambda}{2} \gamma_{kk} \gamma_{nn}\,, \qquad (1.5.25)$$

$$\sigma_{ij} = 2\mu \gamma_{(ij)} + \lambda \gamma_{kk} \delta_{ij}\,, \qquad \mu_{ij} = 0\,. \qquad (1.5.26)$$

It might be observed that the same result is obtained under the assumption that the rotation vector $\underline{\varphi} = 0$. Material constants γ, ε, β being different from zero, since $\underline{\varphi} = 0$, yield $\varkappa_{ij} = \varkappa_{ji} = 0$.

Let us assume that, in addition to $\alpha = 0$, also $\mu = \lambda = 0$. The internal energy takes, thus, the form

$$U = \gamma \varkappa_{(ij)} \varkappa_{(ij)} + \varepsilon \varkappa_{\langle ij \rangle} \varkappa_{\langle ij \rangle} + \frac{\beta}{2} \varkappa_{kk} \varkappa_{nn} , \tag{1.5.27}$$

$$\sigma_{ij} = 0, \quad \mu_{ij} = 2\gamma \varkappa_{(ij)} + 2\varepsilon \varkappa_{\langle ij \rangle} + \beta \varkappa_{kk} \delta_{ij} . \tag{1.5.28}$$

In the case under consideration, the internal energy depends solely upon the tensor $\varkappa_{ji}$ and, hence, upon φ. Couple-stresses μ_{ji} appear in the body and form an asymmetric tensor, whereas stresses σ_{ji} are equal to zero. Thus, formulae (1.5.27 - 28) concern bodies in which the points (particles) can undergo certain rotations but can not be displaced.

1. 6 Compatibility Equations.

Tensors γ_{ji} and $\varkappa_{ji}$ were defined in the following manner

$$\gamma_{ji} = u_{i,j} - \epsilon_{kji} \varphi_k , \quad \varkappa_{ji} = \varphi_{i,j} , \quad i, j, k = 1, 2, 3. \tag{1.6.1}$$

The components of the displacement vector $\underline{u}$ and the rotation vector $\underline{\varphi}$ can be determined from this equation; to determine the six unknown components all 18 differential equations

(1.6.1) must be used. Functions γ_{ji} and $\varkappa_{ji}$ can not be arbitrarily chosen, since they should obey certain conditions which in the classical theory of elasticity are called the compatibility conditions or the compatibility equations of the strain components. We proceed to derive them here.

Let $\underline{u}^\circ$ and $\underline{\varphi}^\circ$ denote the displacement and rotation at point $P^\circ(\underline{x}^\circ)$, and $\underline{u}'$, $\underline{\varphi}'$ - the displacement and rotation at point $P'(\underline{x}')$ of a simple connected body.

The displacement at P' can be expressed with the aid of a curvilinear integral along a continuous line C connecting points P° and P'

$$u_i'(\underline{x}') = u_i^\circ(\underline{x}^\circ) + \int_{P^\circ}^{P'} d u_i = u_i^\circ(\underline{x}^\circ) + \int_{P^\circ}^{P'} \frac{\partial u_i}{\partial x_j} dx_j . \quad (1.6.2)$$

Similarly, the rotation at P' is expressed by

$$\varphi_i'(\underline{x}') = \varphi_i^\circ(\underline{x}') + \int_{P^\circ}^{P'} d\varphi_i = \varphi_i^\circ(\underline{x}^\circ) + \int_{P^\circ}^{P'} \frac{\partial \varphi_i}{\partial x_j} dx_j . \quad (1.6.3)$$

Taking into account Eq.(1.6.1), formula (1.6.2) can be rewritten

$$u_i'(\underline{x}') = u_i^\circ(\underline{x}^\circ) + \int_{P^\circ}^{P'} (\gamma_{ji} + \epsilon_{kji}\varphi_k) dx_j . \quad (1.6.4)$$

The last term in (1.6.4) is integrated by parts

$$\epsilon_{kji}\int_{P^\circ}^{P'} \varphi_k \, dx_j = \epsilon_{kji}\int_{P^\circ}^{P'} \varphi_k \, d(x_j - x_j') =$$

$$= \epsilon_{kji}(x_j^\circ - x_j')\varphi_k^\circ + \epsilon_{kji}\int_{P^\circ}^{P'} (x_j - x_j') \frac{\partial \varphi_k}{\partial x_l} dx_l . \quad (1.6.5)$$

Here φ_k° is the value of φ_k at point $P^\circ(\underline{x}^\circ)$. dx_j has been replaced in Eq.(1.6.5) by $d(x_j - x_j')$. This is permissible, since P' is determined in the process of integration in such a way that $dx_j' = 0$. From (1.6.4) we obtain, finally,

$$(1.6.6)\qquad u_i'(\underline{x}') = u_i^\circ(\underline{x}^\circ) + \epsilon_{kji}(x_j' - x_j^\circ)\varphi_k^\circ + \int_{P^\circ}^{P'} U_{\ell i}\, dx_\ell \,,$$

where

$$U_{\ell i} = \gamma_{\ell i} - \epsilon_{kji}(x_j - x_j')\,\varkappa_{\ell k}\,.$$

Eq.(1.6.3) yields

$$(1.6.7)\qquad \varphi_i'(\underline{x}') = \varphi_i^\circ(\underline{x}^\circ) + \int_{P^\circ}^{P'} \varkappa_{\ell i}\, dx_\ell \,.$$

The integrals appearing in Eq.(1.6.6),(1.6.7) depend on the end points P° and P' of the path of integration but are independent of the integration path itself, provided $u_i'(\underline{x}')$ and $\varphi_i'(\underline{x}')$ are continuous and single-valued functions. The integrands $U_{\ell i}$, $\varkappa_{\ell i}$ should be total differentials.

The necessary conditions for the integrand functions $U_{\ell i}$, $\varkappa_{\ell i}$ to represent total differentials are

$$(1.6.8)\qquad \frac{\partial U_{\ell i}}{\partial x_h} = \frac{\partial U_{hi}}{\partial x_\ell}\,, \qquad \frac{\partial \varkappa_{\ell i}}{\partial x_h} = \frac{\partial \varkappa_{hi}}{\partial x_\ell}\,.$$

These conditions are simultaneously the sufficient conditions for the integrals (1.6.6 - 7) to be single-valued functions in the simple connected region.

Eq.(1.6.8) yields the relations

$$\gamma_{li,h} - \gamma_{hi,l} - \epsilon_{khi}\varkappa_{lk} + \epsilon_{kli}\varkappa_{hk} = 0\,, \qquad (1.6.9)$$

$$\varkappa_{li,h} = \varkappa_{hi,l}\,, \qquad (1.6.10)$$

which represent the compatibility conditions sought for.

1. 7 Equations of Motion in Terms of Displacements and Rotations.

Let us substitute in the equations of motion

$$\sigma_{ji,j} + X_i - \rho\ddot{u}_i = 0 \qquad (1.7.1)$$

$$\epsilon_{ijk}\sigma_{jk} + \mu_{ji,j} + Y_i - \mathcal{J}\ddot{\varphi}_i = 0 \qquad (1.7.2)$$

the stresses

$$\sigma_{ji} = (\mu+\alpha)\gamma_{ji} + (\mu-\alpha)\gamma_{ij} + \lambda\gamma_{kk}\delta_{ij}\,, \qquad (1.7.3)$$

$$\mu_{ji} = (\gamma+\varepsilon)\varkappa_{ji} + (\gamma-\varepsilon)\varkappa_{ij} + \beta\varkappa_{kk}\delta_{ji}\,, \qquad (1.7.4)$$

and use the relations

$$\gamma_{ji} = u_{i,j} - \epsilon_{kji}\varphi_k\,, \quad \varkappa_{ji} = \varphi_{i,j}\,. \qquad (1.7.5)$$

The following system of equations in displacements and rotations results from these substitutions

$$\Box_2\underline{u} + (\lambda+\mu-\alpha)\,\mathrm{grad\,div}\,\underline{u} + 2\alpha\,\mathrm{rot}\,\underline{\varphi} + \underline{X} = 0\,, \qquad (1.7.6)$$

$$\Box_4\underline{\varphi} + (\beta+\gamma-\varepsilon)\,\mathrm{grad\,div}\,\underline{\varphi} + 2\alpha\,\mathrm{rot}\,\underline{u} + \underline{Y} = 0\,. \qquad (1.7.7)$$

Here the differential operators have been introduced

$$\Box_2 = (\mu + \alpha)\nabla^2 - \rho\partial_t^2 , \quad \Box_4 = (\gamma + \varepsilon)\nabla^2 - 4\alpha - \mathcal{J}\partial_t^2 .$$

The hyperbolic equations of elastokinetics are to be completed by the proper set of boundary and initial conditions.

In the case when displacements $\underline{U}(\underline{x}, t)$ and rotations $\underline{\Omega}(\underline{x},t)$ are prescribed on the surface A bounding the body, the boundary conditions assume the form

$$(1.7.8) \qquad \underline{u}(\underline{x}, t) = \underline{U}(\underline{x}, t), \quad \underline{\varphi}(\underline{x}, t) = \underline{\Omega}(\underline{x}, t), \quad \underline{x} \in A .$$

In the case of prescribed values of the surface forces $\hat{\underline{p}}(\underline{x},t)$ and couples $\hat{\underline{m}}(\underline{x},t)$, the boundary conditions are written as

$$(1.7.9) \quad \sigma_{ji}(\underline{x}, t) n_j(\underline{x}) = \hat{p}_i(\underline{x}, t) , \quad \mu_{ji}(\underline{x}, t) n_j(\underline{x}) = \hat{m}_i(\underline{x}, t), \quad \underline{x} \in A .$$

Here $\underline{n}$ is the unit normal vector to the surface A .

Mixed boundary conditions are also possible when on a part A_u of the boundary A displacements and rotations are prescribed, and on the remaining part A_σ - forces and couples are given $(A = A_u + A_\sigma)$.

The initial conditions are assumed in the form

$$(1.7.10) \qquad \begin{aligned} \underline{u}(\underline{x}, 0) &= \underline{g}(\underline{x}), & \dot{\underline{u}}(\underline{x}, 0) &= \underline{f}(\underline{x}), \\ \underline{\varphi}(\underline{x}, 0) &= \underline{k}(\underline{x}), & \dot{\underline{\varphi}}(\underline{x}, 0) &= \underline{h}(\underline{x}), \quad \underline{x} \in V. \end{aligned}$$

Equations (1.7.6) and (1.7.7) form a hyperbolic system of mutually coupled differential equations; they can be uncoupled by the assumption $\alpha = 0$, that leads to two independent systems of equations

$$\mu\nabla^2\underline{u} + (\lambda+\mu)\,\mathrm{grad\,div}\,\underline{u} + \underline{X} = \rho\ddot{u}\,, \qquad (1.7.11)$$

$$(\gamma+\varepsilon)\nabla^2\underline{\varphi} + (\gamma+\beta-\varepsilon)\,\mathrm{grad\,div}\,\underline{\varphi} + \underline{Y} = \mathcal{J}\ddot{\underline{\varphi}}\,. \qquad (1.7.12)$$

Eqs.(1.7.11) are easily recognized as the equations of classical elastokinetics, whereas Eqs.(1.7.12) concern the hypothetical elastic medium in which only rotations are possible. From (1.7.3) and (1.7.4) it is observed that for $\alpha = 0$ the stress tensor σ_{ji} becomes symmetric, the couple stress tensor μ_{ji} remains, however, asymmetric.

Let us return to the system of coupled equations (1.7.6 - 7). Applying the divergence operator, the system is split into two independent equations

$$\left(\nabla^2 - \frac{1}{c_1^2}\,\partial_t^2\right)\gamma_{kk} + \frac{1}{c_1^2\rho}\,\mathrm{div}\,\underline{X} = 0, \qquad (1.7.13)$$

$$\left(\nabla^2 - \frac{1}{c_3^2}\,\partial_t^2 - \delta_o^2\right)\varkappa_{kk} + \frac{1}{\mathcal{J}c_3^2}\,\mathrm{div}\,\underline{Y} = 0\,, \qquad (1.7.14)$$

where

$$\gamma_{kk} = \mathrm{div}\,\underline{u}\,, \qquad \varkappa_{kk} = \mathrm{div}\,\underline{\varphi}$$

and

$$c_1 = \left(\frac{\lambda+2\mu}{\rho}\right)^{1/2}, \quad c_3 = \left(\frac{\beta+2\gamma}{\mathcal{J}}\right)^{1/2}, \quad \delta_o^2 = \frac{4\alpha}{\mathcal{J}c_3^2}\,.$$

Equation (1.7.13) describes the propagation of a

dilatational wave in an elastic space and is identical with the wave equation of the classical elastokinetics.

The other wave equation (1.7.14) describing the propagation of a torsional (rotational) wave has no classical counterpart in traditional elastokinetics. The structure of Eq. (1.7.14) is more complex than that of Eq. (1.7.13); equations of this type are encountered in the problems of quantum electrodynamics and bear the name of the Klein-Gordon differential equation.

In an infinite space the dilatational wave can be produced by div $\underline{X}$, which plays the role of a source in Eq.(1.7.13). Similarly the source of the rotational waves is provided by the function div $\underline{Y}$. Owing to the relations

$$\sigma_{kk} = 3K\gamma_{kk}\ , \quad \mu_{kk} = 3L\varkappa_{kk}$$

it is seen that the first invariants of stress satisfy the wave equations: invariant σ_{kk} - the first equation (1.7.13), and invariant μ_{kk} - the other equation (1.7.14).

Apply now the operation of curl to Eqs.(1.7.6) and (1.7.7) to obtain the system

$$\Box_2 \underline{S} + 2\alpha \operatorname{rot} \underline{T} + \operatorname{rot} \underline{X} = 0\ , \tag{1.7.15}$$

$$\Box_4 \underline{T} + 2\alpha \operatorname{rot} \underline{S} + \operatorname{rot} \underline{Y} = 0 \tag{1.7.16}$$

where

$$\underline{S} = \operatorname{rot} \underline{u}\ , \quad \underline{T} = \operatorname{rot} \underline{\varphi}\ .$$

These equations are coupled with each other. $\underline{S}$ and $\underline{T}$ depend here on the distribution of the body forces $\underline{X}$ and body couples $\underline{Y}$.

Resolving the system with respect to $\underline{S}$ and $\underline{T}$ we are led to

$$\left(\Box_2\Box_4 + 4\alpha^2\nabla^2\right)\underline{S} = 2\alpha \,\mathrm{rot\,rot}\,\underline{Y} - \Box_4 \mathrm{rot}\,\underline{X} \,, \quad (1.7.17)$$

$$\left(\Box_2\Box_4 + 4\alpha^2\nabla^2\right)\underline{T} = 2\alpha \,\mathrm{rot\,rot}\,\underline{X} - \Box_2 \mathrm{rot}\,\underline{Y} \,. \quad (1.7.18)$$

Let us return to Eqs.(1.7.15),(1.7.16) and substitute $\alpha = 0$, to obtain the system

$$\left(\mu\nabla^2 - \rho\partial_t^2\right)\underline{S} + \mathrm{rot}\,\underline{X} = 0 \,, \quad (1.7.19)$$

$$\left((\gamma + \varepsilon)\nabla^2 - J\partial_t^2\right)\underline{T} + \mathrm{rot}\,\underline{Y} = 0 \,. \quad (1.7.20)$$

Equation (1.7.19) is identical with the wave equation characterizing the propagation of the quantity $\mathrm{rot}\,\underline{u}$ in classical elastokinetics. Eq.(1.7.20) describes the propagation of the quantity $\underline{\varphi}$ in a hypothetical medium able to perform rotations only. In an infinite elastic space the sources of disturbances are provided, in accordance with Eqs.(1.7.17),(1.7.18), by the quantities $\mathrm{rot}\,\underline{X}$ and $\mathrm{rot}\,\underline{Y}$.

In a micropolar medium the quantity $\underline{S} = \mathrm{rot}\,\underline{u}$ can be produced either by $\mathrm{rot}\,\underline{X}$ or by $\mathrm{rot}\,\underline{Y}$.

1. 8 Potentials and Stress Functions.

Consider the equations of motion in terms of displacements and rotations

$$(1.8.1)\quad \square_2 \underline{u} + (\lambda + \mu - \alpha)\,\mathrm{grad}\,\mathrm{div}\,\underline{u} + 2\alpha\,\mathrm{rot}\,\underline{\varphi} + \underline{X} = 0$$

$$(1.8.2)\quad \square_4 \underline{\varphi} + (\beta + \gamma - \varepsilon)\,\mathrm{grad}\,\mathrm{div}\,\underline{\varphi} + 2\alpha\,\mathrm{rot}\,\underline{u} + \underline{Y} = 0 .$$

The system is coupled and, moreover, rather complicated and incovenient to deal with; hence, our prime objective will be to uncouple it. Two methods of separation are given below.

The first method is analogous to Lamé's procedure of the classical elasticity and consists in the decomposition of vectors $\underline{u}$, $\underline{\varphi}$ into the potential and solenoidal part+), ++).

$$(1.8.3)\quad \underline{u} = \mathrm{grad}\,\Phi + \mathrm{rot}\,\underline{\Psi}\,,\quad \mathrm{div}\,\underline{\Psi} = 0,$$

$$(1.8.4)\quad \underline{\varphi} = \mathrm{grad}\,\Sigma + \mathrm{rot}\,\underline{H}\,,\quad \mathrm{div}\,\underline{H} = 0.$$

Similar decomposition is performed on vectors

$$(1.8.5)\quad \underline{X} = \rho\left(\mathrm{grad}\,\vartheta + \mathrm{rot}\,\underline{\chi}\right),\quad \mathrm{div}\,\underline{\chi} = 0,$$

$$(1.8.6)\quad \underline{Y} = \mathcal{J}\left(\mathrm{grad}\,\sigma + \mathrm{rot}\,\underline{\eta}\right),\quad \mathrm{div}\,\underline{\eta} = 0.$$

+) Lamé G.: Leçons sur la théorie mathématique de l'élasticité des corps solides. Paris, Bachelier, 1852.

++) Palmov W.A.: Fundamental equations of non-symmetric elasticity (in Russian), Prikl.Mat.Mech. 28 (1964), p.401.

By inserting the above expressions in Eqs.(1.8.1 - 2) we are led to the following system of wave equations

$$\Box_1 \Phi + \rho \vartheta = 0, \tag{1.8.7}$$

$$\Box_3 \Sigma + \mathfrak{I} \sigma = 0, \tag{1.8.8}$$

$$\Box_2 \underline{\Psi} + 2\alpha \operatorname{rot} \underline{H} + \rho \underline{\chi} = 0, \tag{1.8.9}$$

$$\Box_4 \underline{H} + 2\alpha \operatorname{rot} \underline{\Psi} + \mathfrak{I} \underline{\eta} = 0. \tag{1.8.10}$$

Further elimination transforms the last two equations to the form

$$\Omega \underline{\Psi} = 2\alpha \mathfrak{I} \operatorname{rot} \underline{\eta} - \rho \Box_4 \underline{\chi}, \quad \Omega \underline{H} = 2\alpha \rho \operatorname{rot} \underline{\chi} - \mathfrak{I} \Box_2 \underline{\eta}. \tag{1.8.11}$$

The following notations have been introduced here

$$\Box_1 = (\lambda + 2\mu)\nabla^2 - \rho \partial_t^2, \quad \Box_3 = (\beta + 2\gamma)\nabla^2 - 4\alpha - \mathfrak{I}\partial_t^2,$$

$$\Omega = \Box_2 \Box_4 + 4\alpha^2 \nabla^2.$$

The complex system of hyperbolic euqations (1.8.1 - 2) has been reduced to the solution of simple wave equations (1.8.7 - 8) and Eqs.(1.8.11). Eq.(1.8.7) represents a longitudinal wave, Eq. (1.8.8) - a rotational wave; Eqs.(1.8.11) correspond to a transversal-torsional wave.

If $\alpha = 0$ is assumed in Eqs.(1.8.9),(1.8.10), we are led to simple wave equations

$$\left(\mu \nabla^2 - \rho \partial_t^2\right) \underline{\Psi} + \rho \underline{\chi} = 0, \tag{1.8.12}$$

and

$$\left((\gamma+\varepsilon)\nabla^2-\mathcal{J}\partial_t^2\right)\underline{H}+\mathcal{J}\underline{\eta}=0. \tag{1.8.13}$$

It is observed that (1.8.12) is analogous to the transversal wave equation of classical elastokinetics. Eq.(1.8.13) is related to the hypothetical displacement-free medium and can be connected with the solenoidal part of the vector $\underline{Y}$.

Eqs.(1.8.9 - 10) are the wave equations connecting the both types of waves. From (1.8.11) it is evident that quantity $\underline{\Psi}$ can be provoked by $\underline{\chi}$ or $\underline{\eta}$. A similar conclusion applies to the vector $\underline{H}$.

The form of Eq.(1.8.7) is identical with the form of the longitudinal wave equation of classical elastokinetics, contrary to Eq.(1.8.8) which is new; it is a Klein-Gordon differential equation describing the propagation of the potential part of $\underline{\varphi}$.

In the case when the propagation of waves in an unbounded region is considered, the boundary conditions in the strict sense do not appear in the problem. Assuming that the body forces and couples occupy a finite region, functions $\underline{u}$, $\underline{\varphi}$, can be required to tend to zero at infinity. In an infinite space, the particular integrals of Eqs.(1.8.7 - 8) and (1.8.11) constitute the final solutions to the problem. Functions $\Phi, \Sigma, \underline{\Psi}, \underline{H}$ determined, vectors $\underline{u}$ and $\underline{\varphi}$ are found from formulae (1.8.3) and (1.8.4).

In a finite, limited body the waves originating

from the place of disturbance reach the boundary and are reflected. Functions Φ, Σ, $\underline{\Psi}$, $\underline{H}$ are connected with each other by means of the boundary conditions.

The second method of separation of the system (1.8.1 - 2) is analogous to that applied by Galerkin +) to the classical elastostatics, and by Iacovache ++) - to the classical elastokinetics. Functions of this type suitable for asymmetric elasticity ware established by N. Sandru +++) who applied the general algorithm devised by Gr. C. Moisil *).

Another method **) of determining the stress functions will be given here; in our opinion the method is rather simple and makes it possible to avoid the tedious solutions of the sixth order determinants.

Eliminating from Eqs.(1.8.1 - 2) the function $\underline{\varphi}$ and, in turn,

---

+) Galerkin, B.: "Contributions à la solution générale du problème de la théorie de l'élasticité dans le cas de trois dimension", C.R.Acad. Sci. Paris, 190 (1930), p.1047.

++) Iacovache, M.: "O extindere a metodei lui Galerkin pentru sistemul ecuaţiilor elasticităţii", Bull. Ştiint. Acad. Rep. Pop. Române, Ser. A. 1 (1949), 593.

+++) Sandru, N.: "On some problems of the linear theory of asymmetric elasticity" Int. J. Engng. Sci. 4, 1, (1966), 81.

*) Moisil, Gr. C.: "Aspura sistemelor de equaţii cu derivata parţiale lineare şi cu coeficienţi constanti". Bull.Ştiint.Acad. Rep. Pop. Române, Ser. A., (1949), p. 341.

**) Nowacki, W.: "On the completeness of stress functions in asymmetric elasticity", Bull. Acad. Polon. Sci., Sér. Sci. Techn. 14, No 7 (1968), p.309.

the other function $\underline{u}$, we are led to the system

(1.8.14) $$\Omega \underline{u} + \text{grad div}\, \Gamma \underline{u} + \Box_4 \underline{X} - 2\alpha \,\text{rot}\, \underline{Y} = 0 \,,$$

(1.8.15) $$\Omega \underline{\varphi} + \text{grad div}\, \Theta \underline{\varphi} + \Box_2 \underline{Y} - 2\alpha \,\text{rot}\, \underline{X} = 0 \,.$$

where

$$\Omega = \Box_2 \Box_4 + 4\alpha^2 \nabla^2 \,, \qquad \Gamma = (\lambda + \mu - \alpha)\Box_4 - 4\alpha^2 \,,$$

$$\Theta = (\beta + \gamma - \varepsilon)\Box_2 - 4\alpha \,.$$

Let us consider the system of equations (1.8.14) which can be rewritten in the operator form

(1.8.16) $$L_{ij}(u_j) + \Box_4 X_i - 2\alpha \epsilon_{ijk} Y_{k,j} = 0 \,, \quad i,j,k = 1,2,3$$

where

$$L_{ij} = \Omega \delta_{ij} + \partial_i \partial_j \Gamma \,.$$

Introduce the vector function $\underline{\xi}$ connected with the displacement components $\underline{u}$ by means of the relations

(1.8.17) $$u_1 = \begin{vmatrix} \xi_1 & L_{12} & L_{13} \\ \xi_2 & L_{22} & L_{23} \\ \xi_3 & L_{32} & L_{33} \end{vmatrix} , \quad u_2 = \begin{vmatrix} L_{11} & \xi_1 & L_{13} \\ L_{21} & \xi_2 & L_{23} \\ L_{31} & \xi_3 & L_{33} \end{vmatrix} , \quad u_3 = \begin{vmatrix} L_{11} & L_{12} & \xi_1 \\ L_{21} & L_{22} & \xi_2 \\ L_{31} & L_{32} & \xi_3 \end{vmatrix} .$$

After simple transformations the following relation between the vectors $\underline{u}$ and $\underline{\xi}$ is found to be

(1.8.18) $$\underline{u} = \Box_1 \Box_4 \underline{\xi} - \text{grad div}\, \Gamma \underline{\xi} .$$

Applying an analogous procedure to Eq.(1.8.15), we are led to

$$\underline{\varphi} = \Box_2 \Box_3 \underline{\lambda} - \operatorname{grad} \operatorname{div} \Theta \underline{\lambda}, \tag{1.8.19}$$

where $\underline{\lambda}$ is a second vector stress function.

Substituting (1.8.18) into (1.8.14) and (1.8.19) into (1.8.15), the system of equations is derived

$$\Omega \Box_1 \Box_4 \underline{\xi} = -\Box_4 \underline{X} + 2\alpha \operatorname{rot} \underline{Y}, \tag{1.8.20}$$

$$\Omega \Box_2 \Box_3 \underline{\lambda} = -\Box_2 \underline{Y} + 2\alpha \operatorname{rot} \underline{X}. \tag{1.8.21}$$

from which functions $\underline{\xi}$ and $\underline{\lambda}$ can be determined.

Hence, two independent systems of equations have been found; they are not very convenient for further considerations due to the differential operators applied to the right--hand side body forces and couples. If, however, the representations (1.8.18 - 19) are replaced by

$$\underline{u} = \Box_1 \Box_4 \underline{F} - \operatorname{grad} \operatorname{div} \Gamma \underline{F} - 2\alpha \operatorname{rot} \Box_3 \underline{M}, \tag{1.8.22}$$

$$\underline{\varphi} = \Box_2 \Box_3 \underline{M} - \operatorname{grad} \operatorname{div} \Theta \underline{M} - 2\alpha \operatorname{rot} \Box_1 \underline{F} \tag{1.8.23}$$

where $\underline{F}$ and $\underline{M}$ denote the new stress functions, the substitution of (1.8.22) into (1.8.14) and (1.8.23) into (1.8.15) leads to

$$\Box_4 (\Omega \Box_1 \underline{F} + \underline{X}) - 2\alpha \operatorname{rot} (\Omega \Box_3 \underline{M} + \underline{Y}) = 0,$$

$$\Box_2 (\Omega \Box_3 \underline{M} + \underline{Y}) - 2\alpha \operatorname{rot} (\Omega \Box_1 \underline{F} + \underline{X}) = 0.$$

These relations yield the two equations

(1.8.24) $\square_1 \Omega \underline{F} + \underline{X} = 0\,, \qquad \square_3 \Omega \underline{M} + \underline{Y} = 0\,,$

which serve to determine the stress functions $\underline{F}$ and $\underline{M}$. These equations coincide with those obtained in a different way by N. Sandru+).

Equations (1.8.14) were also derived by J. Stefaniak++) who started from Eqs.(1.8.7) to (1.8.10).

Let us quote, in addition, the relations between the stress functions and potentials Φ, Σ, Ψ, $\underline{H}$. Consider the homogeneous equations (1.8.24), the body forces and couplesbeing absent,

(1.8.25) $\square_1 \Omega \underline{F} = 0\,, \qquad \square_3 \Omega \underline{M} = 0\,.$

The particular solution of these equations can be constructed, according by to the theorem by T. Boggio+++) of two parts

(1.8.26) $\underline{F} = \underline{F}' + \underline{F}''\,, \qquad \underline{M} = \underline{M}' + \underline{M}''\,.$

Functions $\underline{F}'$, $\underline{F}''$, $\underline{M}'$, $\underline{M}''$ satisfy the following equations

(1.8.27) $\square_1 \underline{F}' = 0\,, \qquad \Omega \underline{F}'' = 0\,,$

(1.8.28) $\square_3 \underline{M}' = 0\,, \qquad \Omega \underline{M}'' = 0\,.$

+) N. Sandru, op. cit. p.53.

++) Stefaniak, J.:"A generalization of Galerkin's functions for asymmetric thermoelasticity", Bull.Acad. Polon.Sci., Série Sci. Tech., 8, 14, (1968), p.391.

+++) Boggio,T.:"Sull'integrazione di alcune equazioni lineari alle derivate parziali", Ann. Mat., Ser.III (1903), 181.

Substituting (1.8.26) into (1.8.22) and (1.8.23) and making use of Eqs.(1.8.27 - 28), the following representation is obtained

$$\underline{u} = \Box_1 \Box_4 \underline{F}'' - \text{grad}\,\text{div}\,\Gamma\left(\underline{F}' + \underline{F}''\right) - 2\alpha\,\text{rot}\,\Box_3 \underline{M}'', \quad (1.8.29)$$

$$\underline{\varphi} = \Box_2 \Box_3 \underline{M}'' - \text{grad}\,\text{div}\,\Theta\left(\underline{M}' + \underline{M}''\right) - 2\alpha\,\text{rot}\,\Box_1 \underline{F}''. \quad (1.8.30)$$

On utilizing the relations

$$\text{rot}\,\text{rot}\,\underline{U} = \text{grad}\,\text{div}\,\underline{U} - \nabla^2 \underline{U},$$

and taking into account

$$\Omega = \Box_2 \Box_4 + 4\alpha^2 \nabla^2 = \Box_1 \Box_4 - \nabla^2 \Gamma = \Box_2 \Box_3 - \Theta \nabla^2,$$

representation (1.8.29 - 30) is reduced to

$$\underline{u} = -\text{grad}\,\text{div}\,\Gamma \underline{F}' - 2\alpha\,\text{rot}\,\Box_3 \underline{M}'' - \text{rot}\,\text{rot}\,\Gamma \underline{F}'', \quad (1.8.31)$$

$$\underline{\varphi} = -\text{grad}\,\text{div}\,\Theta \underline{M}' - 2\alpha\,\text{rot}\,\Box_1 \underline{F}'' - \text{rot}\,\text{rot}\,\Theta \underline{M}''. \quad (1.8.32)$$

Comparison of the Helmholtz representation (1.8.3) (1.8.4) with the representation (1.8.31 - 32) proceeds to

$$\Phi = -\text{div}\,\Gamma \underline{F}', \quad \underline{\Psi} = -2\alpha\,\Box_3 \underline{M}'' - \text{rot}\,\Gamma \underline{F}'', \quad (1.8.33)$$

$$\Sigma = -\text{div}\,\Theta \underline{M}', \quad \underline{H} = 2\alpha\,\Box_1 \underline{F}'' - \text{rot}\,\Theta \underline{M}''. \quad (1.8.34)$$

These are the relations between potentials Φ, Σ, $\underline{\Psi}$, $\underline{H}$ and the stress functions $\underline{F}$, $\underline{M}$ sought for.

It should be verified here if Eqs.(1.8.33 - 34) satisfy the homogeneous wave equations (1.8.7),(1.8.8) and (1.8.11); this can be easily done and the conclusion is positive.

1. 9 The Principle of Virtual Work.

The validity of the following equation is easily verified,

$$\int_V \left[(X_i - \rho \ddot{u}_i)\,\delta u_i + (Y_i - J\ddot{\varphi}_i)\,\delta\varphi_i\right] dV + \int_A (p_i\,\delta u_i + m_i\,\delta\varphi_i)\, dA =$$

(1.9.1)
$$= \int_V (\sigma_{ji}\,\delta\gamma_{ji} + \mu_{ji}\,\delta\varkappa_{ji})\, dV .$$

The left-hand side represents the virtual work done by the external forces and inertia forces, the right-hand side - the virtual work done by the internal forces. Quantities δu_i and $\delta\varphi_i$ are the virtual increments of displacements u_i and rotations φ_i. Quantities δu_i, $\delta\varphi_i$ are assumed to be infinitesimal and arbitrary; they represent continuous functions and do not violate the external constraints.

The simplest way to derive Eq.(1.9.1) begins from the equations of motion

(1.9.2)
$$\sigma_{ji,j} + X_i - \rho\ddot{u}_i = 0 ,$$

(1.9.3)
$$\epsilon_{ijk}\sigma_{jk} + \mu_{ji,j} + Y_i - J\ddot{\varphi}_i = 0 .$$

Multiplying the first equation by δu_i, the second one - by $\delta\varphi_i$, adding the products and integrating over the volume we come to

$$\int_V \left[(X_i - \rho\ddot{u}_i)\,\delta u_i + (Y_i - J\ddot{\varphi}_i)\,\delta\varphi_i\right] dV + \int_V (\sigma_{ji,j}\,\delta u_i +$$
$$+ \epsilon_{ijk}\sigma_{jk}\,\delta\varphi_i + \mu_{ji,j}\,\delta\varphi_i)\, dV = 0 .$$

After suitable transformation of the second integral, Eq.(1.9.1) is obtained. Now, the stress-strain relations are substituted into the right-hand side of (1.9.1)

$$\sigma_{ij} = 2\mu\gamma_{(ij)} + \lambda\gamma_{kk}\delta_{ij} + 2\alpha\gamma_{\langle ij\rangle}, \tag{1.9.4}$$

$$\mu_{ij} = 2\gamma\varkappa_{(ij)} + \beta\varkappa_{kk}\delta_{ij} + 2\varepsilon\varkappa_{\langle ij\rangle}. \tag{1.9.5}$$

which, after some rearrangements, leads to

$$\int_V \left[(X_i - \rho\ddot{u}_i)\delta u_i + (Y_i - J\ddot{\varphi}_i)\delta\varphi_i\right] dV + \\ + \int_A (p_i\delta u_i + m_i\delta\varphi_i)\, dA = \delta\mathcal{W}, \tag{1.9.6}$$

where

$$\mathcal{W} = \mathcal{U} = \int_V \Big(\mu\gamma_{(ij)}\gamma_{(ij)} + \alpha\gamma_{\langle ij\rangle}\gamma_{\langle ij\rangle} + \frac{\lambda}{2}\gamma_{kk}\gamma_{nn} + \\ + \gamma\varkappa_{(ij)}\varkappa_{(ij)} + \varepsilon\varkappa_{\langle ij\rangle}\varkappa_{\langle ij\rangle} + \frac{\beta}{2}\varkappa_{kk}\varkappa_{nn}\Big)\, dV.$$

Were the couple stresses neglected, Eq.(1.9.6) would express the virtual work principle of the classical elastokinetics,

$$\int_V (X_i - \rho\ddot{u}_i)\delta u_i\, dV + \int_A p_i\delta u_i\, dA = \delta\mathcal{W}_\varepsilon, \\ \mathcal{W} = \int_V \Big(\mu\varepsilon_{ij}\varepsilon_{ij} + \frac{\lambda}{2}\varepsilon_{kk}\varepsilon_{nn}\Big)\, dV, \quad \varepsilon_{ij} = \gamma_{(ij)}. \tag{1.9.7}$$

Passing to the particular case let us assume that the deformations are provoked by causes which are harmonic in time,

$$X_i(\underline{x},t)=e^{-i\omega t}X_i^*(\underline{x}),\quad p_i(\underline{x},t)=e^{-i\omega t}p_i^*(\underline{x}).$$

Here ω is the angular frequency of vibrations. The field of displacements $u_i(\underline{x},t)=e^{-i\omega t}u_i^*(\underline{x})$ and rotations $\varphi_i(\underline{x},t)=$ $=e^{-i\omega t}\varphi_i^*(\underline{x})$ is produced within the body. It is assumed that displacements and rotations are prescribed on the part A_u of the surface A bounding the body, forces and couples being prescribed on the remaining part A_σ of surface A. Hence,

$$\begin{aligned} &p_i(\underline{x},t)=e^{-i\omega t}p_i^*(\underline{x})=\sigma_{ji}^*(\underline{x})n_j(\underline{x})e^{-i\omega t}, \\ (1.9.9)\quad & \\ &m_i(\underline{x},t)=e^{-i\omega t}m_i(\underline{x})=\mu_{ji}^*(\underline{x})n_j(\underline{x})e^{-i\omega t},\quad \underline{x}\in A_\sigma \end{aligned}$$

and

$$(1.9.10)\quad u_i(\underline{x},t)=e^{-i\omega t}f_i(\underline{x}),\quad \varphi_i(\underline{x},t)=e^{-i\omega t}g_i(\underline{x}),\quad \underline{x}\in A_u.$$

The equations of motion take the form

$$(1.9.11)\qquad \sigma_{ji,j}^*+X_i+\rho\omega^2u_i^*=0,$$

$$(1.9.12)\qquad \epsilon_{ijk}\sigma_{jk}^*+\mu_{ji,j}^*+Y_i^*+J\omega^2\varphi_i^*=0.$$

Introduce, now, the virtual amplitudes of displacements δu_i^* and rotation $\delta\varphi_i^*$.

Multiplying Eq.(1.9.11) by δu_i^*, Eq.(1.9.12) - by $\delta\varphi_i^*$, adding the results and performing the integration over the region V , the following equation is obtained

$$\int_V \left[\left(X_i^* + \rho\omega^2 u_i^*\right)\delta u_i^* + \left(Y_i^* + \mathcal{J}\omega^2\varphi_i^*\right)\delta\varphi_i^*\right] dV +$$

$$+ \int_A \left(p_i^*\delta u_i^* + m_i^*\delta\varphi_i^*\right) dA = \int_V \left(\sigma_{ji}^*\delta\gamma_{ji}^* + \mu_{ji}^*\delta\varkappa_{ji}^*\right) dV. \qquad (1.9.13)$$

Substitution of the amplitudes σ_{ji}^* and μ_{ji}^* from (1.9.4 - 5) into (1.9.13) leads to

$$\int_V \left[\left(X_i^* + \rho\omega^2 u_i^*\right)\delta u_i^* + \left(Y_i^* + \mathcal{J}\omega^2\varphi_i^*\right)\delta\varphi_i^*\right] dV +$$

$$+ \int_A \left(p_i^*\delta u_i^* + m_i^*\delta\varphi_i^*\right) dA = d\mathcal{W}^*, \qquad (1.9.14)$$

where

$$\mathcal{W}^* = \int_V \left(\mu\gamma_{(ij)}^*\gamma_{(ij)}^* + \alpha\gamma_{\langle ij\rangle}^*\gamma_{\langle ij\rangle}^* + \frac{\lambda}{2}\gamma_{kk}^*\gamma_{nn}^* + \right.$$

$$\left. + \gamma\varkappa_{(ij)}^*\varkappa_{(ij)}^* + \varepsilon\varkappa_{\langle ij\rangle}^*\varkappa_{\langle ij\rangle}^* + \frac{\beta}{2}\varkappa_{kk}^*\varkappa_{nn}^*\right) dV.$$

Additional notation

$$\mathcal{K}^* = \frac{\omega^2}{2}\int_V \left(\rho u_i^* u_i^* + \mathcal{J}\varphi_i^*\varphi_i^*\right) dV,$$

reduces Eq. (1.9.14) to the form

$$\delta\left[\mathcal{W}^* - \mathcal{K}^* - \int_V \left(X_i^* u_i^* + Y_i^*\varphi_i^*\right) dV - \int_{A_\sigma} \left(p_i^* u_i^* + m_i^*\varphi_i^*\right) dA\right] = 0. \qquad (1.9.15)$$

We are faced here with a certain extremum property of the expression in brackets. It can be seen that this is a problem of a minimum value of the functional. A surface integral taken over the region A_σ appears in Eq.(1.9.15) which results from the fact that displacements and rotations were prescribed on A_u, and - according to previous assumptions - their variations should vanish on the surface A_u.

In the limiting case of a static problem when $\omega \to 0$, expression $\mathcal{K}^*$ vanishes and Eq.(1.9.15) transforms into the minimum potential energy theorem for a micropolar continuum.

1.10 Hamilton's Principle.

From the virtual work principle by varying the displacements and rotations, the Hamilton principle generalized to a micropolar body can be derived. Consider an elastic body changing its position continuously between two instants: $t = t_1$ and $t = t_2$.

Let us compare the actual displacements $\underline{u}(\underline{x}, t)$ and rotations $\underline{\varphi}(\underline{x}, t)$ with displacements $\underline{u} + \delta\underline{u}$ and rotations $\underline{\varphi} + \delta\underline{\varphi}$, variations $\delta\underline{u}$, $\delta\underline{\varphi}$ being selected to satisfy

$$\delta\underline{u}(\underline{x}, t_1) = \delta\underline{u}(\underline{x}, t_2) = 0, \quad \delta\underline{\varphi}(\underline{x}, t_1) = \delta\underline{\varphi}(\underline{x}, t_2) = 0. \tag{1.10.1}$$

Apply the principle of virtual work written in the form

$$\delta L - \int_V (\rho \ddot{u}_i \delta u_i + \mathcal{J} \ddot{\varphi}_i \delta \varphi_i) dV = \delta \mathcal{W}$$

where

$$\delta L = \int_V (X_i \delta u_i + Y_i \delta \varphi_i) dV + \int_A (p_i \delta u_i + m_i \delta \varphi_i) dA. \tag{1.10.2}$$

Here δL is the virtual work done by external forces, and δW represents the virtual work of internal forces given by formula (1.9.6). On integrating Eq.(1.10.2) over the time interval

$$\delta \int_{t_1}^{t_2} W\,dt = \int_{t_1}^{t_2} \delta L\,dt - \int_{t_1}^{t_2}\int_V \left(\rho \ddot{u}_i \delta u_i + J \ddot{\varphi}_i \delta \varphi_i\right) dV\,dt \,, \qquad (1.10.3)$$

and introducing the kinetic energy

$$K = \frac{1}{2}\int_V \left(\rho \dot{u}_i \dot{u}_i + J \dot{\varphi}_i \dot{\varphi}_i\right) dV \,, \qquad (1.10.4)$$

its variation can be written in the form

$$\delta K = \rho \int_V \frac{\partial}{\partial t}\left(\dot{u}_i \delta u_i\right) dV - \rho \int_V \ddot{u}_i \delta u_i\,dV +$$
$$+ J \int_V \frac{\partial}{\partial t}\left(\dot{\varphi}_i \delta \varphi_i\right) dV - J \int_V \ddot{\varphi}_i \delta \varphi_i\,dV. \qquad (1.10.5)$$

Integration over the time interval $t_1 \leqslant t \leqslant t_2$, conditions (1.10.1) imposed upon the virtual displacements and rotations being taken into account, yields

$$\delta \int_{t_1}^{t_2} K\,dt = -\rho \int_{t_1}^{t_2} dt \int_V \ddot{u}_i \delta u_i\,dV - J \int_{t_1}^{t_2} dt \int_V \ddot{\varphi}_i \delta \varphi_i\,dV. \qquad (1.10.6)$$

The right-hand side of this expression is identical with the last right-hand integral of Eq.(1.10.3), and Eq.(1.10.3) is transformed to

$$\delta \int_{t_1}^{t_2} \left(W - K\right) dt = \int_{t_1}^{t_2} \delta L\,dt. \qquad (1.10.7)$$

This is Hamilton's principle generalized to micropolar media. The symbol of variation can be interchanged with the integration sign only in the case of conservative external loads having a potential.

In such a case

$$\delta L = -\left(\frac{\partial \mathcal{V}}{\partial u_i}\delta u_i + \frac{\partial \mathcal{V}}{\partial \varphi_i}\delta\varphi_i\right) = -\delta\left(\frac{\partial \mathcal{V}}{\partial u_i}u_i + \frac{\partial \mathcal{V}}{\partial \varphi_i}\varphi_i\right),$$

where $\mathcal{V}$ is the potential of external forces, and Eq.(1.10.7) is transformed to

$$(1.10.8) \qquad \delta\int_{t_1}^{t_2}(\Pi - \mathcal{K})\,dt = 0 \ , \quad \Pi = \mathcal{W} + \mathcal{V} .$$

Π denotes the total potential energy, $\Pi - \mathcal{K}$ is the Lagrangian.The time integral of the Lagrangian function assumes an extremum value in the interval $t_1 \leq t < t_2$. The special advantage of Hamilton's principle consists in its independence of the coordinate system. Returning to Eqs.(1.10.7),(1.10.8) it may be observed, that the potential of external forces $\mathcal{V}$ exists if these loads are independent of the displacements and rotations. This condition is not satisfied in many practical cases, like - for instance - aerodynamic loads acting on aeroplane wings, where the loads depend on the deformations and, sometimes, even upon their time rates of change. In such cases, when the potential of external forces does not exist, Hamilton's principle in the original form (1.10.7) should be applied.

1. 11 Uniqueness of Solutions.

It will be proved now that the solutions of the

fundamental differential equations of a micropolar body are unique. Evidently, it has to be assumed that such solutions exist.

Let us consider a simple connected body deformed under the action of external forces. Forces $\underline{p}$ and couples $\underline{m}$ are prescribed over the surface A_σ, displacements $\underline{u}$ and rotations $\underline{\varphi}$ are prescribed over A_u. Let us assume first that the solutions are not unique and two different solutions $\underline{u}'$, $\underline{\varphi}'$ and $\underline{u}''$, $\underline{\varphi}''$ exist. These solutions have to satisfy the equations of motion

$$\sigma'_{ji,j} + X_i = \rho \ddot{u}'_i \,, \quad \epsilon_{ijk}\sigma'_{jk} + \mu'_{ji,j} + Y_i = \Im \ddot{\varphi}''_i \,, \tag{1.11.1}$$

$$\sigma''_{ji,j} + X_i = \rho \ddot{u}''_i \,, \quad \epsilon_{ijk}\sigma_{jk} + \mu''_{ji,j} + Y_i = \Im \ddot{\varphi}''_i \,. \tag{1.11.2}$$

The following notations are introduced

$$\hat{u}_i = u'_i - u''_i \,, \quad \hat{\varphi}_i = \varphi'_i - \varphi''_i \,, \quad \hat{\sigma}_{ji} = \sigma'_{ji} - \sigma''_{ji} \,, \; \ldots\ldots \tag{1.11.3}$$

Subtracting Eqs.(1.11.1) and (1.11.2) we are led to a homogeneous system of equations

$$\hat{\sigma}_{ji,j} - \rho \ddot{\hat{u}}_i = 0 \,, \quad \epsilon_{ijk}\hat{\sigma}_{jk} + \hat{\mu}_{ji,j} - \Im \ddot{\hat{\varphi}}_i = 0 \,. \tag{1.11.4}$$

The boundary and initial conditions to be satisfied by these quantities are also homogeneous, namely

$$\begin{aligned} &\hat{u}_i(\underline{x}, t) = 0, \quad \hat{\varphi}_i(\underline{x}, t) = 0, \quad \underline{x} \in A_u \,, \\ &\hat{p}_i(\underline{x}, t) = 0, \quad \hat{m}_i(\underline{x}, t) = 0, \quad \underline{x} \in A_\sigma \,, \end{aligned} \tag{1.11.5}$$

and

$$\begin{aligned} &\hat{u}_i(\underline{x}, 0) = 0 \,, \quad \dot{\hat{u}}_i(\underline{x}, 0) = 0 \,, \\ &\hat{\varphi}_i(\underline{x}, 0) = 0 \,, \quad \dot{\hat{\varphi}}_i(\underline{x}, 0) = 0 \,, \quad \underline{x} \in V, \; t = 0 \,. \end{aligned} \tag{1.11.6}$$

Hence, the displacement and rotation field is described by the equations of motion (1.11.4), body forces and couples being absent and by the homogeneous boundary conditions (1.11.5) and the homogeneous initial conditions (1.11.6)

It has to be verified if any deformations can appear in the interior of the body. It is to this end that the virtual work principle is used Eq.(1.9.6):

$$(1.11.7)\quad \int_V \left[(X_i - \rho\ddot{u}_i)\delta u_i + (Y_i - \mathcal{J}\ddot{\varphi}_i)\delta\varphi\right] dV + \int_A (p_i \delta u_i + m_i \delta\varphi_i)\, dA = \delta \mathcal{W}.$$

Let us compare displacement u_i in a point $\underline{x}$ at instant t with the actual displacement which will appear in this point after a lapse of dt. In this case

$$(1.11.8)\quad \delta u_i = \frac{\partial u_i}{\partial t} dt = v_i\, dt.$$

Similarly,

$$(1.11.9)\quad \delta\varphi_i = \frac{\partial \varphi_i}{\partial t} dt = w_i\, dt, \qquad \delta \mathcal{W} = \frac{\partial \mathcal{W}}{\partial t} dt = \dot{\mathcal{W}} dt.$$

By substituting (1.11.8) and (1.11.9) into (1.11.7) and introducing the kinetic energy $\mathcal{K}$ and its time derivatives

$$(1.11.10)\quad \mathcal{K} = \frac{1}{2}\int_V (\rho v_i v_i + \mathcal{J} w_i w_i)\, dV, \quad \dot{\mathcal{K}} = \int_V (\rho v_i \dot{v}_i + \mathcal{J} w_i \dot{w}_i)\, dV,$$

we obtain

$$(1.11.11)\quad \frac{d}{dt}(\mathcal{W} + \mathcal{K}) = \int_V (X_i v_i + Y_i w_i)\, dV + \int_A (p_i v_i + m_i w_i)\, dA.$$

This equation, called the fundamental energy equation, is used to prove the uniqueness theorem. Adapting Eq.(1.11.11) for the state $\hat{u}_i$, $\hat{\varphi}_i$, $\hat{\sigma}_{ji}$, $\hat{\mu}_{ji}$ ect. it is seen that the right-hand side equals zero because in the interior of the body $\hat{X}_i = \hat{Y}_i = 0$, and on the surface $\hat{u}_i = 0$, $\hat{\varphi}_i = 0$ over A_u, whereas $\hat{p}_i = 0$. $\hat{m}_i = 0$ over A_σ. We are left with the equation

$$\frac{d}{dt}\left(\hat{\mathcal{W}} + \hat{\mathcal{K}}\right) = 0, \tag{1.11.12}$$

whence

$$\hat{\mathcal{W}} + \hat{\mathcal{K}} = \text{const.} \tag{1.11.13}$$

This constant must be zero owing to the initial equations (1.11.6) It is known, however, that the kinetic energy and strain energy cannot be negative; vanishing of their sum means that both terms of (1.11.13) must vanish separately.

The kinetic energy is zero when $\hat{v}_i = 0$, $\hat{w}_i = 0$ at every point $\underline{x}$ and instant t, which leads to the conclusion

$$u_i' = u_i'', \quad \varphi_i' = \varphi_i'', \tag{1.11.14}$$

proving that the displacements and rotations are unique. By equallingto zero the work of deformation which constitutes the quadraticform of deformations yields

$$\gamma_{ji}' = \gamma_{ji}'', \quad \varkappa_{ji}' = \varkappa_{ji}''. \tag{1.11.15}$$

From the constitutive equations follows the conclusion that

stresses

(1.11.16) $$\sigma'_{ji} = \sigma''_{ji}\ ,\quad \mu'_{ji} = \mu''_{ji}\ ,$$

are also unique.

1. 12 Reciprocity of Works Theorem.

This theorem - one of the most fundamental ones - can serve as a starting point for deducing the method of integration of the differential equations of micropolar elasticity. Let us derive a certain identity from the constitutive equations.

Let the body be subject to two systems of causes producing two different systems of effects. External forces, displacements and rotations prescribed over the boundary of the body are considered as causes. One of these two systems of causes and effects will be marked by "primes".

Constitutive equations for the two systems are

(1.12.1) $$\sigma_{ji} = (\mu+\alpha)\gamma_{ji} + (\mu-\alpha)\gamma_{ij} + \lambda\gamma_{kk}\delta_{ji}\ ,$$

(1.12.2) $$\mu_{ji} = (\gamma+\varepsilon)\varkappa_{ji} + (\gamma-\varepsilon)\varkappa_{ij} + \beta\varkappa_{kk}\delta_{ji}\ ,$$

and

(1.12.3) $$\sigma'_{ji} = (\mu+\alpha)\gamma'_{ji} + (\mu-\alpha)\gamma'_{ij} + \lambda\gamma'_{kk}\delta_{ji}\ ,$$

(1.12.4) $$\mu'_{ji} = (\gamma+\varepsilon)\varkappa'_{ji} + (\gamma-\varepsilon)\varkappa'_{ij} + \beta\varkappa'_{kk}\delta_{ji}\ .$$

Eq.(1.12.1) is now multiplied by γ'_{ji} , and Eq.(1.12.3) - by γ_{ji} , and the results are substracted.

Then

$$\sigma_{ji}\gamma'_{ji} = \sigma'_{ji}\gamma_{ji} \ . \qquad (1.12.5)$$

In a similar way, Eqs.(1.12.2) and (1.12.4) lead to

$$\mu_{ji}\varkappa'_{ji} = \mu'_{ji}\varkappa_{ji} \ . \qquad (1.12.6)$$

Let us apply the Laplace-transform to Eqs.(1.12.1)÷(1.12.4). With the notations

$$\bar{\sigma}_{ji}(\underline{x}, p) = \int_0^t \sigma_{ji}(\underline{x}, t)\, e^{-pt} dt, \quad \bar{\mu}_{ji}(\underline{x}, t) = \int_0^t \mu_{ji}(\underline{x}, t)\, e^{-pt} dt, \ \text{etc.} \qquad (1.12.7)$$

and performing analogous operations as before we obtain

$$\bar{\sigma}_{ji}\bar{\gamma}'_{ji} = \bar{\sigma}'_{ji}\bar{\gamma}_{ji}\,, \qquad \bar{\mu}_{ji}\bar{\varkappa}'_{ji} = \bar{\mu}'_{ji}\bar{\varkappa}_{ji} \ . \qquad (1.12.8)$$

Adding the first and the second equation (1.12.8) together integrating over V we are led to

$$\int_V \left(\bar{\sigma}_{ji}\bar{\gamma}'_{ji} + \bar{\mu}_{ji}\bar{\varkappa}'_{ji}\right) dV = \int_V \left(\bar{\sigma}'_{ji}\bar{\gamma}_{ji} + \bar{\mu}'_{ji}\bar{\varkappa}_{ji}\right) dV \ . \qquad (1.12.9)$$

This is the required identity. It can be transformed with the aid of the equations of motion. Applying the Laplace-transform and assuming homogeneous boundary conditions we have

$$\bar{\sigma}_{ji,j} + \bar{X}_i = \rho p^2 \bar{u}_i\,, \quad \epsilon_{ijk}\bar{\sigma}_{jk} + \bar{\mu}_{ji,j} + \bar{Y}_i = \mathcal{J} p^2 \bar{\varphi}\,, \qquad (1.12.10)$$

and

$$\bar{\sigma}'_{ji,j} + \bar{X}'_i = \rho p^2 \bar{u}'_i\,, \quad \epsilon_{ijk}\bar{\sigma}'_{jk} + \bar{\mu}'_{ji,j} + \bar{Y}'_i = \mathcal{J} p^2 \bar{\varphi}' . \qquad (1.12.11)$$

The transform

$$\bar{\gamma}_{ji} = \bar{u}_{i,j} - \epsilon_{kji}\bar{\varphi}_k \ , \quad \bar{\gamma}'_{ji} = \bar{u}'_{i,j} - \epsilon_{kji}\bar{\varphi}'_k \ . \tag{1.12.12'}$$

and

$$\bar{\varkappa}_{ji} = \bar{\varphi}_{i,j} \qquad , \qquad \bar{\varkappa}'_{ji} = \bar{\varphi}'_{i,j} \ . \tag{1.12.12''}$$

will be also used. Let us transform the integral

$$\int_V \bar{\sigma}_{ji}\bar{\gamma}'_{ji}\, dV = \int_V \bar{\sigma}_{ji}\left(\bar{u}'_{i,j} - \epsilon_{kji}\bar{\varphi}'_k\right) dV =$$
$$= \int_V \left[\left(\bar{\sigma}_{ji}\bar{u}'_i\right)_{,j} - \bar{\sigma}_{ji,j}\bar{u}'_i - \epsilon_{kji}\bar{\varphi}'_k\bar{\sigma}_{ji}\right] dV =$$
$$= \int_A \bar{p}_i\bar{u}'_i\, dA + \int_V \bar{X}_i\bar{u}'_i\, dV - \rho p^2\int_V \bar{u}_i\bar{u}'_i\, dV - \int_V \epsilon_{kji}\bar{\sigma}_{ji}\bar{\varphi}'_k\, dV.$$

Similar calculations give

$$\int_V \bar{\mu}_{ji}\bar{\varkappa}'_{ji}\, dV = \int_V \bar{\mu}_{ji}\bar{\varphi}'_{i,j}\, dV = \int_V \left[\left(\bar{\mu}_{ji}\bar{\varphi}'_i\right)_{,j} - \bar{\mu}_{ji,j}\bar{\varphi}'_i\right] dV =$$
$$= \int_A \bar{m}_i\bar{\varphi}'_i\, dA + \int_V \left(\bar{Y}_i\bar{\varphi}'_i + \epsilon_{ijk}\bar{\sigma}_{jk}\bar{\varphi}'_i\right) dV - \mathcal{J}p^2\int_V \bar{\varphi}_i\bar{\varphi}'_i\, dV \ .$$

After simple rearrangements Eq.(1.12.9) is reduced to the form

$$\int_V \left(\bar{X}_i\bar{u}'_i + \bar{Y}_i\bar{\varphi}'_i\right) dV + \int_A \left(\bar{p}_i\bar{u}'_i + \bar{m}_i\bar{\varphi}'\right) dA =$$
$$= \int_V \left(\bar{X}'_i\bar{u}_i + \bar{Y}'_i\bar{\varphi}_i\right) dV + \int_A \left(\bar{p}'_i\bar{u}_i + \bar{m}'_i\bar{\varphi}\right) dA \ . \tag{1.12.13}$$

The inverse Laplace-transform has to be applied to this expression, using the convolution theorem. As a result, the final form of the reciprocal theorem is obtained

$$\int_V (X_i * u_i' + Y_i * \varphi_i')\, dV + \int_A (p_i * u_i' + m_i * \varphi')\, dA =$$

$$= \int_V (X_i' * u_i + Y_i' * \varphi_i)\, dV + \int_A (p_i' * u_i + m_i' * \varphi)\, dA, \quad (1.12.14)$$

where

$$X_i * u_i' = \int_0^t X_i(\underline{x}, \tau)\, u_i'(\underline{x}, t-\tau)\, d\tau = \int_0^t X_i(\underline{x}, t-\tau)\, u_i'(\underline{x}, \tau)\, d\tau, \text{ etc.}$$

In the case of an infinite body the reciprocal theorem takes a particularly simple form. Let us assume the body forces and couples to act within a finite region Γ. With increasing distances from this region the displacements and rotations decrease and tend to zero, provided $|x_1^2 + x_2^2 + x_3^2| \to \infty$. In such a case Eq.(1.12.14) is reduced to a single term

$$\int_V (X_i * u_i' + Y_i * \varphi_i')\, dV = \int_V (X_i' * u_i + Y_i' * \varphi_i)\, dV, \quad (1.12.15)$$

the integral extending over the entire infinite region.

Let a unit instantaneous concentrated force $X_i = \delta(\underline{x} - \underline{\xi})\, \delta(t)\, \delta_{ij}$ (parallel to x_j-axis) be applied to a point $\underline{\xi}$ of the body. The force produces the field of displacements $U_i^{(j)}(\underline{x}, \underline{\xi}, t)$ and rotations $\Omega_i^{(j)}(\underline{x}, \underline{\xi}, t)$ in the infinite region. At point $\underline{\eta}$ another instantaneous concentrated force $X_i' = \delta(\underline{x} - \underline{\eta})\, \delta(t)\, \delta_{ik}$ can be applied, parallel to the x_k-axis. The fields of displacements and rotation produced by force X_i' is denoted by $u_i' = U_i^{(k)}(\underline{x}, \underline{\eta}, t)$, $\varphi_i' = \Omega_i^{(k)}(\underline{x}, \underline{\eta}, t)$, respectively.

Taking into account the relations

$$\int_0^t \delta(\underline{x}-\underline{\xi})\delta(\tau)\delta_{ij}U_i^{(k)}(\underline{x},\underline{\eta},t-\tau)\,d\tau = \delta(\underline{x}-\underline{\xi})U_j^{(k)}(\underline{x},\underline{\eta},t)$$

$$\int_V \delta(\underline{x}-\underline{\xi})U_j^{(k)}(\underline{x},\underline{\eta},t)\,dV(\underline{x}) = U_j^{(k)}(\underline{\xi},\underline{\eta},t)$$

the following equation results from Eq. (1.12.15)

$$U_j^{(k)}(\underline{\xi},\underline{\eta},t) = U_k^{(j)}(\underline{\eta},\underline{\xi},t), \quad j,k=1,2,3. \tag{1.12.16}$$

Now, a couple $Y_i' = \delta(\underline{x}-\underline{\eta})\delta_{ik}\delta(t)$ acting parallel to the x_k-axis is applied at point $\underline{\eta}$; the corresponding displacements are denoted by $V_i^{(k)}(\underline{x},\underline{\eta},t)$. A force $X_i = \delta(\underline{x}-\underline{\xi})\delta(t)\delta_{ij}$ acts, as before at the point $\underline{\xi}$ producing the field of rotations $\Omega_i^{(j)}(\underline{x},\underline{\xi},t)$. Equation (1.12.15) with $Y_i = 0$, $X_i' = 0$ leads to

$$\int_V dV(\underline{x})\int_0^t \delta(\underline{x}-\underline{\xi})\delta(\tau)\delta_{ij}V_i^{(k)}(\underline{x},\underline{\eta},t-\tau)\,d\tau -$$
$$-\int_V dV(\underline{x})\int_0^t \delta(\underline{x}-\underline{\eta})\delta(\tau)\delta_{ik}\Omega_i^{(j)}(\underline{x},\underline{\xi},t-\tau)\,d\tau = 0,$$

whence, it follows that

$$V_j^{(k)}(\underline{\xi},\underline{\eta},t) = \Omega_k^{(j)}(\underline{\eta},\underline{\xi},t). \tag{1.12.17}$$

We apply now at $\underline{\xi}$ a concentrated, instantaneous couple $Y_i = \delta(\underline{x} - \underline{\xi})\delta(t)\delta_{ij}$, and at $\underline{\eta}$ - another couple $Y'_i = \delta(\underline{x} - \underline{\eta})\delta(t)\delta_{ik}$. The field of rotations produced by couple Y_i is denoted by $\Omega_i^{(j)}(\underline{x}, \underline{\xi}, t)$, and the other field - by $\Omega_i^{(k)}(\underline{x}, \underline{\eta}, t)$. From Eq.(1.12.15) it follows that

$$\Omega_j^{(k)}(\underline{\xi}, \underline{\eta}, t) = \Omega_k^{(j)}(\underline{\eta}, \underline{\xi}, t) . \tag{1.12.18}$$

Relations (1.12.16) through (1.12.18) are to be treated as a generalization of the J.C. Maxwell reciprocal theorem known from the classical elastokinetics. It is easily verified that relations (1.12.16) - (1.12.18) remain true also in the case of a limited body which is rigidly clamped over the surface A_u and stress-free over A_σ.

Let us return to an infinite body and consider the action of a concentrated force $X_i = \delta(x_1)\delta(x_2)\delta(x_3 - vt)$; the force moves along the x_3-axis at a constant velocity v. The corresponding displacements are denoted by $u_i(\underline{x}, t)$, and rotations - by $\varphi_i(\underline{x}, t)$.

The second system of loads is formed by another instantaneous concentrated force $X'_i = \delta(\underline{x} - \underline{\xi})\delta(t)\delta_{ij}$ applied to point $\underline{\xi}$, parallel to the x_j-axis, and provoking the displacements and rotations denoted by $u'_i(\underline{x}, \underline{\xi}, t)$ and $\varphi'_i(\underline{x}, \underline{\xi}, t)$, respectively.

These forces are introduced into Eq.(1.12.15); $Y_i = 0$ and $Y'_i = 0$ are assumed to vanish.

We are led to

$$\int_V dV(\underline{x})\int_0^t\left[\delta(x_1)\delta(x_2)\delta(x_3-v\tau)\delta_{ij}u_i'(\underline{x},\underline{\xi},t-\tau)-\delta(\underline{x}-\underline{\xi})\delta(\tau)\delta_{ij}u_i(\underline{x},t-\tau)\right]d\tau=0$$

whence,

$$u_j(\xi_1,\xi_2,\xi_3,t)=\int_0^t u_j'(0,0,v\tau;\xi_1,\xi_2,\xi_3;t-\tau)\,d\tau. \tag{1.12.19}$$

Applying a similar procedure another formula can be established

$$\varphi_j(\xi_1,\xi_2,\xi_3,t)=\int_0^t \hat{u}_j'(0,0,v\tau;\xi_1,\xi_2,\xi_3;t-\tau)\,d\tau. \tag{1.12.20}$$

In this formula $\varphi_j(\underline{x},t)$ denotes the rotation produced by the force $X_i=\delta(x_1)\,\delta(x_2)\,\delta(x_3-vt)$, and $\hat{u}_j(\underline{x},\underline{\xi},t)$ denotes the displacement produced by the instantaneous, concentrated couple $Y_i'=\delta(\underline{x}-\underline{\xi})\,\delta(t)\,\delta_{ij}$ applied to the point $\underline{\xi}$ and parallel to the x_j-axis.

Let us consider the action of causes varying harmonically with time. It is easily found that the reciprocal theorem takes here the form

$$\int_V\left(X_i^*u_i^{\prime *}+Y_i^*\varphi_i^{\prime *}\right)dV+\int_A\left(p_i^*u_i^{\prime *}+m_i^*\varphi_i^{\prime *}\right)dA=$$

$$=\int_V\left(X_i^{\prime *}u_i^*+Y_i^{\prime *}\varphi_i^*\right)dV+\int_A\left(p_i^{\prime *}u_i^*+m_i^{\prime *}\varphi_i^*\right)dA. \tag{1.12.21}$$

Asterisks denote the amplitudes of the corresponding quantities. Relations (1.12.16) - (1.12.18) remain true here, under the obvious assumption that - for instance -

$$U_j^{(k)}(\underline{\xi},\underline{\eta},t) = U_j^{*(k)}(\underline{\xi},\underline{\eta})\, e^{-i\omega t} .$$

A certain variant of the reciprocal theorem will be derived here; it is assumed that the "primed" state of loading refers to the static problem what enables us to apply the much simpler fundamental solutions, namely the Green functions of the static problem

The following operations are performed. The first equation of motion

$$\sigma_{ji,j} + X_i - \rho \ddot{u}_i = 0 , \tag{1.12.22}$$

is multiplied by u_i' , and the second equation of motion

$$\epsilon_{ijk}\sigma_{jk} + \mu_{ji,j} + Y_i - \mathcal{J}\ddot{\varphi}_i = 0 \tag{1.12.23}$$

by φ_i' . The results are added together and integrated over the volume V . The following equation results

$$\int_V \left[(X_i - \rho\ddot{u}_i)\, u_i' + (Y_i - \mathcal{J}\ddot{\varphi}_i)\, \varphi_i' \right] dV +$$
$$+ \int_V \left(\sigma_{ji,j} u_i' + \epsilon_{ijk}\sigma_{jk}\varphi_i' + \mu_{ji,j}\varphi_i' \right) dV = 0 ,$$

whence, after transformations, we are led to

$$\int_V \left[(X_i - \rho\ddot{u}_i) u_i' + (Y_i - \mathcal{J}\ddot{\varphi}_i)\varphi_i' \right] dV +$$
$$+ \int_A (p_i u_i' + m_i \varphi_i')\, dA = \int_V (\sigma_{ji}\gamma_{ji}' + \mu_{ji}\varkappa_{ji}')\, dV . \tag{1.12.24}$$

Consider now the equation of equilibrium for the "primed" system of loads

$$\sigma'_{ji,j} + X'_i = 0\,, \quad \epsilon_{ijk}\sigma'_{jk} + \mu'_{ji,j} + Y'_i = 0\,. \tag{1.12.25}$$

Multiplying the first equation of (1.12.25) by u_i and the second by φ_i, adding and integrating we obtain - after simple transformations

$$\int_V (X'_i u_i + Y'_i \varphi_i)\, dV + \int_A (p'_i u_i + m'_i \varphi_i)\, dA = \int_V (\sigma'_{ji}\gamma_{ji} + \mu'_{ji}\varkappa_{ji})\, dV\,. \tag{1.12.26}$$

With the aid of the constitutive equations the validity of the relation can be proved

$$\sigma_{ji}\gamma'_{ji} + \mu_{ji}\varkappa'_{ji} = \sigma'_{ji}\gamma_{ji} + \mu'_{ji}\varkappa_{ji}\,. \tag{1.12.27}$$

Identity (1.12.27) renders the left-hand sides of Eqs.(1.12.24) and (1.12.26) equal and

$$\int_V \left[(X_i - \rho\ddot{u}_i)u'_i + (Y_i - J\ddot{\varphi}_i)\varphi'_i\right] dV + \int_A (p_i u'_i + m_i \varphi'_i)\, dA = \int_V (X'_i u_i + Y'_i \varphi_i)\, dV + \int_A (p'_i u_i + m'_i \varphi_i)\, dA\,. \tag{1.12.28}$$

Formula (1.12.28) can be applied to the following problem:

Determine the displacement $u_i(\underline{\xi}, t)$ as the point $\underline{\xi}$ of an infinite elastic space, produced by the action of body forces $X_i(\underline{x}, t)$ and body couples $Y_i(\underline{x}, t)$ distributed within a finite region Γ. $U_i^{(j)}(\underline{x}, \underline{\xi})$ and $\Omega_i^{(j)}(\underline{x}, \underline{\xi})$ denote the displacement and rotation produced by a static concentrated force $X_i = \delta(\underline{x} - \underline{\xi})\delta_{ij}$ applied to the point $\underline{\xi}$ parallel to the x_j-axis. From Eq.(1.12.28), surface integrals being neglected, one obtains

$$u_j(\underline{\xi}, t) + \rho \int_V U_i^{(j)} \ddot{u}_i \, dV + \mathcal{J} \int_V \Omega_i^{(j)} \ddot{\varphi}_i \, dV =$$

$$= \int_V \left(X_i U_i^{(j)} + Y_i \Omega_i^{(j)} \right) dV . \quad (1.12.29)$$

In an analogous manner rotation $\varphi_j(\underline{x}, t)$ is obtained

$$\varphi_j(\underline{\xi}, t) + \rho \int_V \Omega_j^{(i)} \ddot{u}_i \, dV + \mathcal{J} \int_V W_i^{(j)} \ddot{\varphi}_i \, dV =$$

$$= \int_V \left(X_i \Omega_j^{(i)} + Y_i W_i^{(j)} \right) dV . \quad (1.12.30)$$

Here $W_i^{(j)}$ denotes the rotation at $\underline{\xi}$ produced by a static concentrated couple $Y_i' = \delta(\underline{x} - \underline{\xi})\delta_{ij}$ Equations (1.12.29), (1.12.30) represent the solution to the problem and form a system of two integro-differential equations.

If the displacements and rotations are harmonic in time, i.e.

$$u_i(\underline{x}, t) = e^{-i\omega t} u_i^*(\underline{x}) , \quad \varphi_i(\underline{x}, t) = e^{-i\omega t} \varphi_i^*(\underline{x}) , \quad \text{etc.} ,$$

then Eqs.(1.12.29),(1.12.30) are reduced to a system of Fredholm integral equations of the second kind,

$$\varphi_j^*(\underline{\xi}) - \rho\omega^2 \int_V \Omega_j^{(i)} u_i^* \, dV - \mathcal{J}\omega^2 \int_V W_i^{(j)} \varphi_i^* \, dV =$$

(1.12.31)
$$= \int_V \left(X_i^* \Omega_j^{(i)} + Y_i W_i^{(j)} \right) dV ,$$

$$u_j^*(\underline{\xi}) - \rho\omega^2 \int_V U_i^{(j)} u_i^* \, dV - \mathcal{J}\omega^2 \int_V \Omega_i^{(j)} \varphi_i^* \, dV =$$

(1.12.32)
$$= \int_V \left(X_i^* U_i^{(j)} + Y_i^* \Omega_i^{(j)} \right) dV .$$

The formulae derived above remain valid also in the case of finite, bounded bodies that are rigidly clamped over surface A_u and loadfree over surface A_σ . Eqs.(1.12.31),(1.12.32) represent then the forced vibrations of a bounded body. The case of free vibrations is of particular interest and leads to equations

(1.12.33) $$u_j^*(\underline{\xi}) - \rho\omega^2 \int_V U_i^{(j)} u_i^* \, dV - \mathcal{J}\omega^2 \int_V \Omega_i^{(j)} \varphi_i^* \, dV = 0 ,$$

(1.12.34) $$\varphi_j^*(\underline{\xi}) - \rho\omega^2 \int_V \Omega_j^{(i)} u_i^* \, dV - \mathcal{J}\omega^2 \int_V W_i^{(j)} \varphi_i^* \, dV = 0 .$$

From the system of equations the eigenfrequencies $\omega^{(k)}$, $k = 1,2,\ldots,\infty$ can be determined.

The system (1.12.31),(1.12.32) can not be solved for frequencies $\omega^{(1)}$, $\omega^{(2)}$,.... since then we are faced with the resonance.

1. 13 Generalized Somigliana Theorem.

Let us consider the body occupying the volume V bounded by the surface A. Let the body be acted upon by body forces $\underline{X}$ and body couples $\underline{Y}$. On the surface A displacements $\underline{u}$ and rotations $\underline{\varphi}$ are considered as known. The initial conditions are assumed to be homogeneous. The starting point of our considerations is the works reciprocity theorem.

$$\int_V (X_i * u_i' + Y_i * \varphi_i')\,dV + \int_A (p_i * u_i' + m_i * \varphi_i')\,dA =$$
$$= \int_V (X_i' * u_i + Y_i' * \varphi_i)\,dV + \int_A (p_i' * u_i + m_i' * \varphi_i)\,dA, \tag{1.13.1}$$

where

$$X_i * u_i' = \int_0^t X_i(\underline{x},\tau)\,u_i'(\underline{x}, t-\tau)\,d\tau \qquad \text{etc.}$$

Let the forces X_i, couples Y_i the displacements u_i and rotations φ_i be referred to a bounded body, whereas the functions X_i', Y_i', u_i', φ_i' - to an infinite elastic region.

The instantaneous, concentrated force $X_i' = \delta(\underline{x} - \underline{\xi})\,\delta(t)\,\delta_{ik}$ is assumed to act at the point $\underline{\xi} \in V$ of an infinite space; the force is parallel to the X_k -axis and produces the displacement $u_i' = U_i^{(k)}(\underline{x}, \underline{\xi}, t)$ and rotation $\varphi_i' = \Omega_i^{(k)}(\underline{x}, \underline{\xi}, t)$.

Functions $\underline{U}^{(k)}$, $\underline{\Omega}^{(k)}$ are found from the solution (for an infinite space) of the system of equations:

(1.13.2) $$\Box_2 \underline{U}^{(k)} + (\lambda+\mu-\alpha)\,\mathrm{grad\,div}\,\underline{U}^{(k)} + 2\alpha\,\mathrm{rot}\,\underline{\Omega}^{(k)} + \delta(\underline{x}-\underline{\xi})\delta(t)\underline{i}_k = 0,$$

(1.13.3) $$\Box_4 \underline{\Omega}^{(k)} + (\beta+\gamma-\varepsilon)\,\mathrm{grad\,div}\,\underline{\Omega}^{(k)} + 2\alpha\,\mathrm{rot}\,\underline{U}^{(k)} = 0.$$

$\underline{i}_k$ being the unit base vector of the x_k-axis. Homogeneity of the initial conditions has been assumed here

$$\underline{U}^{(k)}(\underline{x},0) = 0, \qquad \underline{\Omega}^{(k)}(\underline{x},0) = 0.$$

Let us now proceed to determine the stresses σ'_{ji}, μ'_{ji} corresponding to the displacements $\underline{U}^{(k)}$ and rotations $\underline{\Omega}^{(k)}$, respectively. They are used to construct the principal force vectors $\underline{p}^{(k)}$ and principal couple vectors $\underline{m}^{(k)}$ on the surface A.

(1.13.4) $$p_i^{(k)} = \sigma'_{ji} n_j, \qquad m_i^{(k)} = \mu'_{ji} n_j$$

By inserting $X'_i = \delta(\underline{x}-\underline{\xi})\,\delta_{ik}\,\delta(t)$, $Y'_i = 0$, $u'_i = U_i^{(k)}$, ... into Eq. (1.13.1) and using the relation

$$\int_0^t d\tau \int_V X'_i(\underline{x},t-\tau)u_i(\underline{x},\tau)dV(\underline{x}) = \int_0^t d\tau \int_V \delta(\underline{x}-\underline{\xi})\delta_{ik}\delta(t-\tau)u_i(\underline{x},\tau)dV(\underline{x}) = u_k(\underline{\xi},t)$$

we finally arrive at

$$u_k(\underline{\xi},t) = \int_V \left(X_i * U_i^{(k)} + Y_i * \Omega_i^{(k)}\right)dV + \int_A \Big(p_i * U_i^{(k)} +$$

(1.13.5) $$+\, m_i * \Omega_i^{(k)} - p_i^{(k)} * u_i - m_i^{(k)} * \varphi_i\Big)dA\,.$$

Let us make the assumption that an instantaneous concentrated couple $Y'_i = \delta(\underline{x}-\underline{\xi})\,\delta(t)\,\delta_{ik}$ acts in the point $\underline{\xi} \in V$ and produces the displacement field $u'_i = V_i^{(k)}(\underline{x},\underline{\xi},t)$

and rotation field $\varphi_i' = \Phi_i^{(k)}(\underline{x}, \underline{\xi}, t)$. The principal vectors on A corresponding to these displacements and rotations are denoted by

$$\hat{p}_i^{(k)} = \sigma_{ji}' n_j , \qquad \hat{m}_i^{(k)} = \mu_{ji}' n_j .$$

Substituting the above quantities into Eq.(1.13.1) we obtain

$$\varphi_k(\underline{\xi}, t) = \int_V \left(X_i * V_i^{(k)} + Y_i * \Phi_i^{(k)} \right) dV + \int_A \left(p_i * V_i^{(k)} + \right.$$

$$\left. + m_i * \Phi_i^{(k)} - \hat{p}_i^{(k)} * u_i - \hat{m}_i^{(k)} * \varphi_i \right) dA . \qquad (1.13.6)$$

Formulae (1.13.5) and (1.13.6) represent the well--known from the classical elastokinetics Somigliana theorem, generalized to asymmetric elasticity. These formulae make it possible to determine the values of $u_k(\underline{\xi}, t)$, $\varphi_k(\underline{\xi}, t)$, $\underline{\xi} \in V$ within the body A once the values of $u_i(\underline{x}, t)$, $\varphi_i(\underline{x}, t)$ and $p_i(\underline{x}, t)$, $m_i(\underline{x}, t)$, $\underline{x} \in A$ on the boundary are known.Eqs.(1.13.5),(1.13.6) hold true as long as $\underline{\xi} \in A$. These formulae, however, are of theoretical meaning only, since either the values of u_i, φ_i or the values of m_i , p_i are usually prescribed on the boundary. If, on the other hand, displacements u_i' and rotations φ_i' are chosen so that functions $U_i^{(k)}$, $\Omega_i^{(k)}$, $V_i^{(k)}$, $\Phi_i^{(k)}$ can be referred to a body with a perfectly clamped surface A , then Eqs.(1.13.5),(1.13.6) will become of practical interest. In order to determine the quantities $U_i^{(k)}$, $\Omega_i^{(k)}$, Equations (1.13.2),(1.13.3) under homogeneous initial conditions should be solved. The latter have the form

$$U_i^{(k)}(\underline{x},\underline{\xi},t)=0, \quad \Omega_i^{(k)}(\underline{x},\underline{\xi},t)=0, \quad \underline{x}\in A.$$

In a similar way the Green functions $V_i^{(k)}$, $\Phi_i^{(k)}$ are determined. With the newly acquired meaning of the functions $U_i^{(k)}$, $\Omega_i^{(k)}$, $V_i^{(k)}$, $\Phi_i^{(k)}$, Eqs.(1.13.5),(1.13.6) (under the assumption that $U_i^{(k)}=0$, $\Omega_i^{(k)}=0$, $V_i^{(k)}=0$, $\Phi_i^{(k)}=0$ on the surface A) take the form

$$(1.13.7)\quad u_k(\underline{\xi},t)=\int_V\left(X_i*U_i^{(k)}+Y_i*\Omega_i^{(k)}\right)dV-\int_A\left(p_i^{(k)}*u_i+m_i^{(k)}*\varphi_i\right)dA,$$

$$(1.13.8)\quad \varphi_k(\underline{\xi},t)=\int_V\left(X_i*V_i^{(k)}+Y_i*\Phi_i^{(k)}\right)dV-\int_A\left(\hat{p}_i^{(k)}*u_i+\hat{m}_i^{(k)}*\varphi_i\right)dA.$$

Formulae (1.13.7) and (1.13.8) furnish the solution of the first boundary value problem of asymmetric elasticity.

In a similar manner, when a body free of loads on its boundary is taken as an auxiliary system, the solution of the second boundary value problem (loads are prescribed on A) can be achieved. We are led to the following expressions

$$(1.13.9)\quad u_k(\underline{\xi},t)=\int_V\left(X_i*u_i'+Y_i*\varphi_i'\right)dV+\int_A\left(p_i*u_i'+m_i*\varphi_i'\right)dA,$$

$$(1.13.10)\quad \varphi_k(\underline{\xi},t)=\int_V\left(X_i*u_i''+Y_i*\varphi_i''\right)dV+\int_A\left(p_i*u_i''+m_i*\varphi_i''\right)dA.$$

Here u_i', φ_i' denote the displacements and rotations produced in the auxiliary system by the force $X_i'=\delta(\underline{x}-\underline{\xi})\delta(t)\delta_{ik}$.

Quantities u_i'', φ_i'' are associated with the action of the couple $Y_i'' = \delta(\underline{x} - \underline{\xi})\delta(t)\delta_{ik}$.

Chapter 2

Special Problems of Elastokinetics.

2. 1 Monochromatic Plane Waves in Elastic Space.

Let us consider a plane wave which varies harmonically in time. The wave front is located at instant t = const. in the plane $p = x_i n_i$, $\underline{n}$ being the unit normal vector to the plane. Thus the displacements and rotations should be assumed in the following form

(2.1.1) $u_j = A_j \exp\left[-ki\left(ct - n_k x_k\right)\right]$

(2.1.2) $\varphi_j = B_j \exp\left[-ki\left(ct - n_k x_k\right)\right], \quad k = \frac{\omega}{c} = \frac{2\pi}{l}, \quad i = \sqrt{-1}$.

Here c is the phase velocity, ω - the angular frequency and l - the wave length.

Introducing (2.1.1),(2.1.2) into the system

(2.1.3) $\square_2 \underline{u} + (\lambda + \mu - \alpha)\,\mathrm{grad\,div}\,\underline{u} + 2\alpha\,\mathrm{rot}\,\underline{\varphi} = 0$,

(2.1.4) $\square_4 \underline{\varphi} + (\beta + \gamma - \varepsilon)\,\mathrm{grad\,div}\,\underline{\varphi} + 2\alpha\,\mathrm{rot}\,\underline{u} = 0$

the following system of six algebraic equations is obtained

(2.1.5) $(\mu + \alpha - \rho c^2)A_j + (\lambda + \mu - \alpha)n_j n_k A_k + \frac{2\alpha i}{k}\epsilon_{jk\ell} n_\ell B_k = 0$,

(2.1.6) $\left(\gamma + \varepsilon + \frac{4\alpha}{k^2} - \Im c^2\right)B_j + (\beta + \gamma - \varepsilon)n_j n_k B_k + \frac{2\alpha i}{k}\epsilon_{jk\ell} n_\ell A = 0$.

The system possesses non-trivial solutions when its determinant vanishes; this condition yields the following equation

$$\left(\lambda+2\mu-\rho c^2\right)\left(2\gamma+\beta+\frac{4\alpha}{k^2}-\mathcal{J}c^2\right)\left[\left(\mu+\alpha-\rho c^2\right)\left(\gamma+\varepsilon+\frac{4\alpha}{k^2}-\mathcal{J}c^2\right)-\frac{4\alpha^2}{k^2}\right]=0, \quad (2.1.7)$$

from which the phase velocities of various types of plane waves will be determined. Namely, from the equation

$$\lambda+2\mu-\rho c^2=0$$

a constant and independent of ω phase velocity is found

$$c=c_1=\left(\frac{\lambda+2\mu}{\rho}\right)^{1/2} . \quad (2.1.8)$$

From the equation

$$2\gamma+\beta+\frac{4\alpha}{k^2}-\mathcal{J}c^2=0$$

another phase velocity results

$$c=c_3\left(1-\frac{\omega_0^2}{\omega^2}\right)^{-1/2}, \quad c_3=\left(\frac{2\gamma+\beta}{\mathcal{J}}\right)^{1/2}, \quad \omega_0^2=\frac{4\alpha}{\mathcal{J}} . \quad (2.1.9)$$

This velocity depends on the frequency ω and belongs to a wave exhibiting dispersion. The phase velocity preserves its meaning only if $\omega>\omega_0$, because then it assumes real values.

Equation

$$\left(\mu+\alpha-\rho c^2\right)\left(\gamma+\varepsilon+\frac{4\alpha}{k^2}-\mathcal{J}c^2\right)-\frac{4\alpha^2}{k^2}=0 \quad (2.1.10)$$

leads to the following equation in k :

$$k^4-k^2\left(\sigma_2^2+\sigma_4^2+p(s-2)\right)+\sigma_2^2\left(\sigma_4^2-2p\right)=0 . \quad (2.1.11)$$

Here the notations have been introduced

$$\sigma_4 = \frac{\omega}{c_4}, \quad \sigma_2 = \frac{\omega}{c_2}, \quad c_4 = \left(\frac{\gamma+\varepsilon}{J}\right)^{1/2}, \quad c_2 = \left(\frac{\mu+\alpha}{\rho}\right)^{1/2},$$

$$s = \frac{2\alpha}{\rho c^2}, \qquad p = \frac{2\alpha}{J c_4^2}.$$

Solutions to the biquadratic equation (2.1.11) are

$$(2.1.12')\quad k_{1,2}^2 = \frac{1}{2}\left(\sigma_2^2 + \sigma_4^2 + p(s-2)\right) \pm \frac{1}{2}\sqrt{\left(\sigma_2^2 + \sigma_4^2 + p(s-2)\right)^2 + 4\sigma_2^2\left(2p - \sigma_4^2\right)},$$

$$(2.1.12'')\quad k_{1,2}^2 = \frac{1}{2}\left(\sigma_2^2 + \sigma_4^2 + p(s-2)\right) \pm \sqrt{\left(\sigma_4^2 - \sigma_2^2 + p(s-2)\right)^2 + 4ps\sigma_2^2}.$$

The form of the last term of the root implies that the discriminant

$$\Delta = \left(\sigma_4^2 - \sigma_2^2 + p(s-2)\right)^2 + 4ps\sigma_2^2$$

is always positive, and quantities k_1^2 and k_2^2 have real values. From (2.1.12) it is seen that for $\sigma_4^2 > 2p$, i.e. for $\omega^2 > \omega_0^2$, the inequalities

$$k_1^2 > 0, \quad k_2^2 > 0,$$

are always true; that corresponds to two real phase velocities. For $\omega^2 < \omega_0^2$ one obtains

$$k_1^2 > 0, \quad k_2^2 < 0.$$

Only one phase velocity is encountered here, the other phase velocity $c = \frac{\omega}{k^2}$ is imaginary and has no physical meaning.

It may be observed that the velocities appearing

in Eq.(2.1.10) depend on the parameter ω ; the waves travelling at these velocities are dispersive.

Let us determine the waves corresponding to phase velocities expressed by formulae (2.1.8),(2.1.9),(2.1.12"). To simplify this procedure let us assume the front of the plane wave to be perpendicular to the x_1-axis. Substituting $n_1 = 1,\ n_2 = n_3 = 0$ into Eqs.(2.1.5),(2.1.6) we are led to a system of six equations

$$\left(\lambda + 2\mu - \rho c^2\right) A_1 = 0 \,, \tag{2.1.13}$$

$$\begin{aligned} &\left(\mu + \alpha - \rho c^2\right) A_2 + \frac{2i\alpha}{k} B_3 = 0 \,, \\ &\left(\gamma + \varepsilon + \frac{4\alpha}{k^2} - \mathcal{J} c^2\right) B_3 - \frac{2i\alpha}{k} A_2 = 0 \,, \end{aligned} \tag{2.1.14}$$

$$\begin{aligned} &\left(2\gamma + \beta + \frac{4\alpha}{k^2} - \mathcal{J} c^2\right) B_1 = 0 \,, \\ &\left(\gamma + \varepsilon + \frac{4\alpha}{k^2} - \mathcal{J} c^2\right) B_2 + \frac{2i\alpha}{k} A_3 = 0 \,, \end{aligned} \tag{2.1.15}$$

$$\left(\mu + \alpha - \rho c^2\right) A_3 - \frac{2i\alpha}{k} B_2 = 0 \,. \tag{2.1.16}$$

It is seen that the phase velocity $c = \left(\dfrac{\lambda + 2\mu}{\rho}\right)^{1/2}$ corresponds to the longitudinal wave

$$u_1(x_1, t) = A_1 \exp\left[-ik\left(ct \mp x_1\right)\right] \,,$$

propagating in the direction of the axis x_1. The velocity $c = c_3\left(1 - \dfrac{\omega_0^2}{\omega^2}\right)^{-1/2}$ is connected with the wave

$$\varphi_1(x_1, t) = B_1 \exp\left[-ik\left(ct \mp x_1\right)\right] \,.$$

This wave is called rotational, and it is associated with the first component of the vector $\underline{\varphi}$. The velocity c resulting from (2.1.12')is associated with two pairs of waves, (u_2, φ_3) or (u_3, φ_2). The first wave corresponds to Eq.(2.1.14), the second one - to Eq.(2.1.16). These waves are called the modified transversal and modified torsional waves.

Let us compare the waves discussed here with analogous plane waves known from the theory of symmetric elasticity. Substituting $\alpha = 0$ and $\gamma = \beta = \varepsilon = 0$ into Eqs.(2.1.5) and (2.1.6) we obtain

$$(\mu - \rho c^2) A_j + (\lambda + \mu) n_j n_k A_k = 0 . \tag{2.1.17}$$

By equating to zero the determinant of this system of equations

$$(2\mu + \lambda - \rho c^2)(\mu - \rho c^2)^2 = 0 ,$$

the phase velocities are obtained

$$c' = \left(\frac{\lambda + 2\mu}{\rho}\right)^{1/2}, \quad c'' = c''' = \left(\frac{\mu}{\rho}\right)^{1/2} . \tag{2.1.18}$$

Velocity c' refers to the longitudinal wave, and its value coincides with c_1 calculated from Eq.(2.1.8). Velocities c'', c''' are associated with the transversal waves u_2, u_3 . They are not dispersive, their phase velocities being independent of ω .

In the micropolar elasticity, quantities $v_1 = \frac{\omega}{k_1}$, $v_2 = \frac{\omega}{k_2}$ represent the counterparts of the phase velocities, c'', c''' and k_1, k_2 being the roots of Equation (2.1.11). Let us finally observe that the wave φ_1 whose phase velocity is expressedby (2.1.9) has no counterpart in the classical elastokinetics.

Introducing $\alpha = 0$ and $A_j = 0$ into (2.1.5) and (2.1.6) the following system of equations is obtained

$$(\gamma + \varepsilon - \mathcal{J}c^2) B_j + (\beta + \gamma - \varepsilon) n_j n_k B_k = 0 . \qquad (2.1.19)$$

By equating to zero the determinant of this system we are led to

$$(\gamma + 2\beta - \mathcal{J}c^2)(\gamma + \varepsilon - \mathcal{J}c^2) = 0 .$$

Here the phase velocity

$$c' = \left(\frac{\gamma + 2\beta}{\mathcal{J}}\right)^{1/2},$$

is associated with the wave φ_1 , and velocities

$$c'' = c''' = \left(\frac{\gamma + \varepsilon}{\mathcal{J}}\right)^{1/2}$$

are associated with the waves φ_1 and φ_2 . These waves can occur only in a hypothetical elastic continuum consisting of particles capable of rotations but unable to be displaced.

Returning to Eqs.(2.1.3),(2.1.4) let us represent them in the particular form corresponding to the propagation of a plane wave in the direction of the x_1-axis. Assuming that all quantities u_i , φ_i depend upon the variables x_1, t only, the following system of equations (2.1.20 - 23) can be written

$$(2.1.20)\quad \left(\partial_1^2 - \frac{1}{c_1^2}\partial_t^2\right) u_1(x_1, t) = 0 \ ,$$

$$(2.1.21)\quad \left(\partial_1^2 - \frac{\nu^2}{c_3^2} - \frac{1}{c_3^2}\partial_t^2\right) \varphi_1(x_1, t) = 0 \ ,$$

$$(2.1.22)\quad \begin{aligned} &\left(\partial_1^2 - \frac{1}{c_2^2}\partial_t^2\right) u_2 - s\,\partial_1 \varphi_3 = 0 \ , \\ &\left(\partial_1^2 - \frac{1}{c_4^2}\partial_t^2 - 2p\right)\varphi_3 + p\,\partial_1 u_2 = 0 \ , \end{aligned}$$

$$(2.1.23)\quad \begin{aligned} &\left(\partial_1^2 - \frac{1}{c_2^2}\partial_t^2\right) u_3 + s\,\partial_1 \varphi_2 = 0 , \\ &\left(\partial_1^2 - \frac{1}{c_4^2}\partial_t^2 - 2p\right)\varphi_2 - p\,\partial_1 u_3 = 0 \ , \qquad \nu^2 = \frac{4\alpha}{J} \ . \end{aligned}$$

Eq.(2.1.20) represents a longitudinal wave, Eq.(2.1.21) - a torsional wave, Eqs.(2.1.22 - 23) - the modified transversal and torsional waves. Owing to the assumed monochromatic character of the waves

$$(2.1.24)\quad u_i(x_1, t) = e^{-i\omega t} u_i^*(x_1) \ , \quad \varphi_i(x_1, t) = e^{-i\omega t} \varphi_i^*(x_1) \ ,$$

Eqs.(2.1.20) through (2.1.23) are reduced to the form

$$(2.1.25)\quad \left(\partial_1^2 + \sigma_1^2\right) u_1^*(x_1) = 0 \ ,$$

$$(2.1.26)\quad \begin{aligned} &\left(\partial_1^2 + \sigma_3^2 - \frac{\nu^2}{c_3^2}\right) \varphi_1^*(x_1) = 0 \ , \\ &\sigma_1 = \frac{\omega}{c_1} \ , \quad \sigma_3 = \frac{\omega}{c_3} \ . \end{aligned}$$

$$\left(\partial_1^2+\sigma_2^2\right)u_2^*-s\,\partial_1\varphi_3^*=0\,,$$
$$\left(\partial_1^2+\sigma_4^2-2p\right)\varphi_3^*+p\,\partial_1 u_2^*=0\,, \qquad (2.1.27)$$

$$\left(\partial_1^2+\sigma_2^2\right)u_3^*+s\,\partial_1\varphi_2^*=0\,,$$
$$\left(\partial_1^2+\sigma_4^2-2p\right)\varphi_2^*-p\,\partial_1 u_3^*=0\,. \qquad (2.1.28)$$

The solution to Eq.(2.1.27) is the function

$$u_1(x_1,t)=U_+^\circ\, e^{-i\omega\left(t-\frac{x_1}{c_1}\right)}+U_-^\circ\, e^{-i\omega\left(t+\frac{x_1}{c_1}\right)}\,. \qquad (2.1.29)$$

This is a longitudinal wave; the first right-hand term represents a wave propagating in the positive direction, the second one - in the negative direction of the x_1 -axis. The wave is undamped and non-dispersive. The solution of Eq.(2.1.26) is the function

$$\varphi_1=A_+\, e^{-i\omega\left(t-\frac{x_1}{c}\right)}+A_-\, e^{-i\omega\left(t+\frac{x_1}{c}\right)}\,,\quad c=c_3\left(1-\frac{\omega_0^2}{\omega^2}\right)^{-1/2}. \qquad (2.1.30)$$

The limitation $\omega>\omega^\circ$ is to be remembered here; the wave is undamped though dispersive.

Substituting into Eq.(2.1.27)

$$u_2^*(x_1)=\overset{\circ}{u}_2\, e^{ikx_1}\,,\quad \varphi_3^*(x_1)=\overset{\circ}{\varphi}_3\, e^{ikx_1}\,,$$

we are led to the following two algebraic equations

$$\overset{\circ}{u}_2\left(\sigma_2^2-k^2\right)-sik\,\overset{\circ}{\varphi}_3=0$$

$$\left(\sigma_4^2-k^2-2p\right)\overset{\circ}{\varphi}_3+pik\,\overset{\circ}{u}_2=0\,,$$

yielding the relations

$$\frac{\overset{\circ}{u}_2}{\overset{\circ}{\varphi}_3} = \frac{sik}{\sigma_2^2 - k^2} = -\frac{\sigma_4^2 - k^2 - 2p}{pik} ,$$

whence

$$\left(\sigma_2^2 - k^2\right)\left(\sigma_4^2 - k^2 - 2p\right) - spk^2 = 0 ;$$

that leads to Eq.(2.1.11).

The solution of the system (2.1.27) is furnished by the functions

$$\text{(2.1.31)} \quad \begin{aligned} u_2 &= B_+ e^{-i\omega\left(t - \frac{x_1}{v_1}\right)} + B_- e^{-i\omega\left(t + \frac{x_1}{v_1}\right)} + \\ &+ \frac{isk_2}{\sigma_2^2 - k_2^2}\left(C_+ e^{-i\omega\left(t - \frac{x_1}{v_2}\right)} - C_- e^{-i\omega\left(t + \frac{x_1}{v_2}\right)}\right), \end{aligned}$$

$$\text{(2.1.32)} \quad \begin{aligned} \varphi_3 &= C_+ e^{-i\omega\left(t - \frac{x_1}{v_2}\right)} - C_- e^{-i\omega\left(t + \frac{x_1}{v_2}\right)} + \\ &+ \frac{\sigma_2^2 - k_1^2}{isk_1}\left(B_+ e^{-i\omega\left(t - \frac{x_1}{v_1}\right)} - B_- e^{-i\omega\left(t + \frac{x_1}{v_1}\right)}\right), \quad v_\beta = \frac{\omega}{k_\beta}, \beta = 1, 2. \end{aligned}$$

These relations represent two types of waves. One pair with the phase velocities v_1, v_2 moves toward the positive, the other - toward the negative direction of the x_1-axis. All wave terms occuring in (2.1.31-32) are dispersive.

The solutions of the system of equations (2.1.28) are analogous to those of Eqs.(2.1.31-32). The stresses σ_{ji}, μ_{ji} associated with the propagation of a plane wave are listed below

$$
\begin{aligned}
&\sigma_{11}=(2\mu+\lambda)\,\partial_1 u_1\,, && \sigma_{22}=\sigma_{33}=\lambda\partial_1 u_1\,,\\
&\sigma_{12}=(\mu+\alpha)\,\partial_1 u_2-2\alpha\varphi_3\,, && \sigma_{21}=(\mu+\alpha)\,\partial_1 u_2+2\alpha\varphi_3\,,\\
&\sigma_{13}=(\mu+\alpha)\,\partial_1 u_3+2\alpha\varphi_2\,, && \sigma_{31}=(\mu+\alpha)\,\partial_1 u_3-2\alpha\varphi_2\,,\\
&\sigma_{23}=-2\alpha\varphi_1\,, && \sigma_{32}=2\alpha\varphi_1\,.
\end{aligned}
\tag{2.1.33}
$$

$$
\begin{aligned}
&\mu_{11}=(2\gamma+\beta)\,\partial_1\varphi_1\,, && \mu_{22}=\mu_{33}=\beta\partial_1\varphi_1\,,\\
&\mu_{12}=(\gamma+\varepsilon)\,\partial_1\varphi_2\,, && \mu_{21}=(\gamma-\varepsilon)\,\partial_1\varphi_2\,,\\
&\mu_{13}=(\gamma+\varepsilon)\,\partial_1\varphi_3\,, && \mu_{31}=(\gamma-\varepsilon)\,\partial_1\varphi_3\,,\\
&\mu_{23}=\mu_{32}=0\,.
\end{aligned}
\tag{2.1.34}
$$

It is to be observed that the normal force stresses σ_{11}, σ_{22}, σ_{33} are associated with the longitudinal wave u_1, and the couple-stresses μ_{11}, μ_{22}, μ_{33} and shearing stresses σ_{23}, σ_{32} - with the microrotational wave φ_1. Finally, stresses σ_{12}, σ_{21}, μ_{13}, μ_{32} are associated with the coupled waves u_2, φ_3 and stresses σ_{31}, σ_{13}, μ_{12}, μ_{21} - with the coupled waves u_3, φ_2. Only the normal stresses σ_{11}, σ_{22}, σ_{33} do not undergo the dispersion.

It has been shown that the system of equations in terms of displacements and rotations (2.1.3 - 4) can be replaced by a system of wave equations, owing to the decomposition of vectors $\underline{u}$ and $\underline{\varphi}$ into the potential and solenoidal parts.

(2.1.35) $$\underline{u} = \text{grad}\,\Phi + \text{rot}\,\underline{\Psi}, \qquad \text{div}\,\underline{\Psi} = 0,$$

(2.1.36) $$\underline{\varphi} = \text{grad}\,\Sigma + \text{rot}\,\underline{H}, \qquad \text{div}\,\underline{H} = 0.$$

The wave equations have the form

(2.1.37) $$\Box_1 \Phi = 0,$$

(2.1.38) $$\Box_3 \Sigma = 0,$$

(2.1.39) $$\Box_2 \underline{\Psi} + 2\alpha\,\text{rot}\,\underline{H} = 0,$$

(2.1.40) $$\Box_4 \underline{H} + 2\alpha\,\text{rot}\,\underline{\Psi} = 0.$$

Let us investigate the behaviour of these equations in the case when a plane wave is propagated. Substituting in Eq.(2.1.37) the expression

(2.1.41) $$\Phi = \overset{\circ}{\Phi} \exp\left[ik(\underline{n}\cdot\underline{r} - ct)\right], \quad \underline{n}\cdot\underline{r} = n_k \cdot x_k$$

describing the propagation of a plane wave in the direction of $\underline{n}$ normal to the plane E, the phase velocity $c_1 = \left(\frac{\lambda + 2\mu}{\rho}\right)^{1/2}$ is obtained, and it is found that

(2.1.42) $$\underline{u}' = \text{grad}\,\Phi = ik_1 \overset{\circ}{\Phi}\,\underline{n} \exp\left[ik_1(\underline{n}\cdot\underline{r} - c_1 t)\right], \quad k_1 = \frac{\omega}{c_1}.$$

Displacement $\underline{u}'$ is directed toward the normal $\underline{n}$. The wave (2.1.42) is called the longitudinal displacement wave.

By substituting into Eq.(2.1.38)

$$\Sigma = \Sigma^{\circ} \exp\left[ik(\underline{n}\cdot\underline{r} - c't)\right] \tag{2.1.43}$$

the phase velocity is obtained

$$c' = \frac{c_3}{\sqrt{1-\frac{\omega_o^2}{\omega^2}}}\ , \quad c_3 = \left(\frac{\beta+2\gamma}{J}\right)^{1/2}, \quad \omega_o^2 = \frac{4\alpha}{J}\ .$$

The rotation $\underline{\varphi}'$ associated with the potential Σ is expressed by Eq.(2.1.36)

$$\underline{\varphi}' = \mathrm{grad}\,\Sigma = ik'\Sigma^{\circ}\underline{n}\cdot\exp\left[ik'(\underline{n}\cdot\underline{r} - c't)\right], \qquad k' = \frac{\omega}{c'}\ . \tag{2.1.44}$$

Here it is also concluded that the direction of vector $\underline{\varphi}'$ coincides with the direction of the normal $\underline{n}$. The wave (2.1.44) is called the longitudinal micro-rotational wave.

Let us substitute into Eq.(2.1.39) and (2.1.40) the expressions

$$\begin{aligned} \underline{\Psi} &= \underline{\Psi}^{\circ}\exp\left[ik(\underline{n}\cdot\underline{r} - ct)\right], \\ \underline{H} &= \underline{H}^{\circ}\exp\left[ik(\underline{n}\cdot\underline{r} - ct)\right]. \end{aligned} \tag{2.1.45}$$

As a result, the coupled system of equations is derived

$$\begin{aligned} \left(\sigma_2^2 - k^2\right)\underline{\Psi}^{\circ} + s\underline{n}\times\underline{H}^{\circ} &= 0\ , \\ \left(\sigma_4^2 - k^2 - \nu^2\right)\underline{H}^{\circ} + p\underline{n}\times\underline{\Psi}^{\circ} &= 0\ . \end{aligned} \tag{2.1.46}$$

Owing to the second Eqs.(2.1.35) and (2.1.36)

$$\underline{n}\cdot\underline{\Psi}^{\circ} = 0\ , \qquad \underline{n}\cdot\underline{H}^{\circ} = 0\ . \tag{2.1.47}$$

Equations (2.1.47) indicate that the vectors $\underline{\Psi}^{\circ}$, $\underline{H}^{\circ}$ lie in a plane perpendicular to the normal $\underline{n}$.

From (2.1.46) it follows that

$$\underline{\Psi}^{\circ} = -\frac{s}{\sigma_2^2 - k^2}\,\underline{n}\times\underline{H}^{\circ},$$

(2.1.48)

$$\underline{H}^{\circ} = -\frac{p}{\sigma_4^2 - k^2 - \nu^2}\,\underline{n}\times\underline{\Psi}^{\circ}.$$

This implies that vectors $\underline{\Psi}^{\circ}$, $\underline{H}^{\circ}$, $\underline{n}$ are mutually perpendicular. Vectors $\underline{\Psi}^{\circ}$, $\underline{H}^{\circ}$ are coupled with each other and cannot exist separately. The direction of propagation of the plane wave $\underline{n}$ and the positions of vectors $\underline{\Psi}$, $\underline{H}$ are shown in Fig.2.1.1.

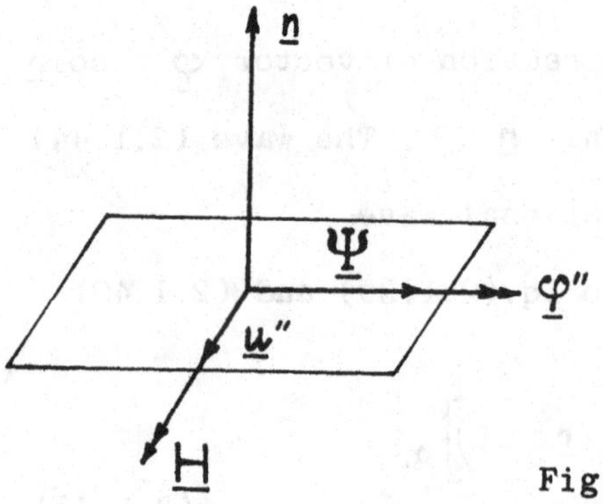

Fig.2.1

Let us determine the quantities $\underline{u}''$, $\underline{\varphi}''$ associated with vectors $\underline{\Psi}^{\circ}$, $\underline{H}^{\circ}$. Eqs.(2.1.35),(2.1.36) yield

(2.1.49) $$\underline{u}'' = \underline{n}\times\underline{\Psi}^{\circ}\exp\left[ik\left(\underline{n}\cdot\underline{r} - ct\right)\right],$$

(2.1.50) $$\underline{\varphi}'' = \underline{n}\times\underline{H}^{\circ}\exp\left[ik\left(\underline{n}\cdot\underline{r} - ct\right)\right].$$

It is evident that vectors $\underline{u}''$, $\underline{\varphi}''$ are coplanar with vectors

$\underline{\Psi}^\circ$, $\underline{H}^\circ$, that is visualized in Fig. 2.1.1.

Waves (2.1.49),(2.1.50) are called transversal; Equation (2.1.49) describes a transversal displacement wave, Eq. (2.1.50) - a transversal micro-rotational wave.

2. 2 Rotational and Longitudinal Waves in an Infinite Elastic Space +)

In Sec. 1.8. the separation of the system of differential equations

$$\Box_2 \underline{u} + (\lambda + \mu - \alpha)\,\mathrm{grad\,div}\,\underline{u} + 2\alpha\,\mathrm{rot}\,\underline{\varphi} + \underline{X} = 0, \qquad (2.2.1)$$

$$\Box_4 \underline{\varphi} + (\beta + \gamma - \varepsilon)\,\mathrm{grad\,div}\,\underline{\varphi} + 2\alpha\,\mathrm{rot}\,\underline{u} + \underline{Y} = 0, \qquad (2.2.2)$$

was performed and, as a result, a system of wave equations was derived. Decomposition of the vectors $\underline{u}$, $\underline{\varphi}$ and $\underline{X}$, $\underline{Y}$ into the potential and rotational parts

$$\underline{u} = \mathrm{grad}\,\Phi + \mathrm{rot}\,\underline{\Psi}\,, \qquad \mathrm{div}\,\underline{\Psi} = 0\,,$$

$$\underline{\varphi} = \mathrm{grad}\,\Sigma + \mathrm{rot}\,\underline{H}\,, \qquad \mathrm{div}\,\underline{H} = 0\,, \qquad (2.2.3)$$

and

$$\underline{X} = \rho\left(\mathrm{grad}\,\vartheta + \mathrm{rot}\,\underline{\chi}\right), \qquad \mathrm{div}\,\underline{\chi} = 0\,,$$

$$\underline{Y} = \mathfrak{J}\left(\mathrm{grad}\,\sigma + \mathrm{rot}\,\underline{\eta}\right), \qquad \mathrm{div}\,\underline{\eta} = 0\,, \qquad (2.2.4)$$

+) W. Nowacki: "Propagation of rotational waves in asymmetric elasticity", Bull. Acad. Polon. Sci., Série Sci. Techn. 16, No 10 (1963) led us to the following wave equations

$$\square_1 \Phi + \rho \vartheta = 0 ,$$

$$\square_3 \Sigma + \mathcal{J} \sigma = 0 ,$$

(2.2.5)

$$\Omega \underline{\Psi} = 2\alpha \mathcal{J} \operatorname{rot} \underline{\eta} - \rho \square_4 \underline{\chi} ,$$

$$\Omega \underline{H} = 2\alpha \rho \operatorname{rot} \underline{\chi} - \mathcal{J} \square_2 \underline{\eta} , \quad \Omega = \square_2 \square_4 + 4\alpha^2 \nabla^2 .$$

A more detailed consideration of the equation governing the rota tional waves is shown here:

(2.2.6) $$\left(\nabla^2 - \frac{1}{c_3^2}\partial_t^2 - \frac{\nu^2}{c_3^2}\right)\Sigma(\underline{x},t) = -\frac{1}{c_3^2}\sigma(\underline{x},t)$$

with $$c_3 = \left(\frac{\beta + 2\gamma}{\mathcal{J}}\right)^{1/2} , \qquad \nu^2 = \frac{4\alpha}{\mathcal{J}} .$$

Equations of this type do not occur in the classical elastokinetics; they are well-known - as the Klein-Gordon equations - from the quantum electrodynamics. It is observed that the first part of Eq.(2.2.5) describing the longitudinal wave propagation

(2.2.7) $$\left(\nabla^2 - \frac{1}{c_1^2}\partial_t^2\right)\Phi(\underline{x},t) + \rho\vartheta(\underline{x},t) = 0 , \quad c_1 = \left(\frac{\lambda + 2\mu}{\rho}\right)^{1/2} ,$$

can be considered as a particular case of Eq.(2.2.6) with ν equal zero. The solutions of the second part of differential equation (2.2.5) and, in parallel, the solutions for the function Φ will be given in the sequel.

To consider the non-homogeneous equation (2.2.6) let us first assume that the initial conditions of the problem are homogeneous. Assume, moreover, the sources $\sigma(\underline{x}, t)$ causing the propagation of the rotational waves to be distributed in a bounded region V'.

For the solution of Eq.(2.2.6) the Green function $G(\underline{x}, \underline{\xi}, t)$ is used, satisfying the differential equation

$$\left(\nabla^2 - \frac{1}{c_3^2}\partial_t^2 - \frac{\nu^2}{c_3^2}\right) G(\underline{x}, \underline{\xi}, t) = -4\pi\delta(\underline{x} - \underline{\xi})\delta(t), \quad (2.2.8)$$

with homogeneous initial conditions

$$G(\underline{x}, \underline{\xi}, 0) = 0, \qquad G'(\underline{x}, \underline{\xi}, 0) = 0, \quad (2.2.9)$$

and under the assumption that $G \to 0$ for $|x_1^2 + x_2^2 + x_3^2| \to \infty$.

The right-hand side of Eq.(2.2.8) represents an instantaneous and concentrated disturbance of intensity 4π, acting at $\underline{\xi}$.

The Laplace-integral-transform is applied to Eq. (2.2.8), initial conditions (2.2.9) being taken into account. One obtains

$$\left(\nabla^2 - \frac{1}{c_3^2}(p^2 + \nu^2)\right) \bar{G}(\underline{x}, \underline{\xi}, p) = -4\pi\delta(\underline{x} - \underline{\xi}), \quad (2.2.10)$$

where

$$\bar{G}(\underline{x}, \underline{\xi}, p) = \int_0^\infty G(\underline{x}, \underline{\xi}, t)\, e^{-pt} dt .$$

The solution of Eq.(2.2.10) is given by the function

$$\bar{G}(\underline{x}, \underline{\xi}, p) = \frac{1}{R}\exp\left[-\frac{R}{c_3}\sqrt{p^2 + \nu^2}\right], \quad R = \left[(\underline{x}_i - \underline{\xi}_i)(\underline{x}_i - \underline{\xi}_i)\right]^{1/2}. \quad (2.2.11)$$

Let us now apply the integral-trasform to Eq.(2.2.6) under the assumption that the initial conditions are homogeneous,

$$\left(\nabla^2 - \frac{1}{c_3^2}\left(p^2 + \nu^2\right)\right)\bar{\Sigma}(\underline{x}, p) = -\frac{1}{c_3^2}\bar{\sigma}(\underline{x}, p)\ . \tag{2.2.12}$$

With the aid of a suitable combination of Eqs.(2.2.10) and (2.2.12), and by integrating over the region V, the relation follows

$$\int_V \left(\bar{G}\nabla^2\bar{\Sigma} - \bar{\Sigma}\nabla^2\bar{G}\right)dV = -\frac{1}{c_3^2}\int_V \bar{\sigma}\bar{G}\,dV + 4\pi\int_V \delta(\underline{x}-\underline{\xi})\bar{\Sigma}(\underline{x}, p)\,dV,$$

which, by means of the Green-transformation, is transformed to

(2.2.13)

$$\bar{\Sigma}(\underline{\xi}, p) = \frac{1}{4\pi c_3^2}\int_V \bar{\sigma}(\underline{x}, p)\bar{G}(\underline{x}, \underline{\xi}, p)\,dV(\underline{x}) + \frac{1}{4\pi}\int_A \left(\bar{G}\frac{\partial\bar{\Sigma}}{\partial n} - \bar{\Sigma}\frac{\partial \bar{G}}{\partial n}\right)dA\ .$$

Here $\frac{\partial}{\partial n}$ denotes the derivative with respect to the outer normal to A . When the propagation of a rotational wave in an infinite elastic region is considered, the surface integral in Eq. (2.2.13) vanishes, and, with the aid of (2.2.11), we finally obtain

$$\bar{\Sigma}(\underline{\xi}, p) = \frac{1}{4\pi c_3^2}\int_{V'} \frac{\bar{\sigma}(\underline{x}, p)}{R(\underline{x}, \underline{\xi})}e^{-\frac{Rp}{c_3}}\,dV(\underline{x}) +$$

$$+\frac{1}{4\pi c_3^2}\int_{V'} \frac{\bar{\sigma}(\underline{x}, p)}{R(\underline{x}, \underline{\xi})}\bar{F}(\underline{x}, \underline{\xi}, p)\,dV(\underline{x})\ , \tag{2.2.14}$$

where

$$\bar{F}(\underline{x}, \underline{\xi}, p) = \exp\left(-\frac{R}{c_3}\sqrt{p^2+\nu^2}\right) - \exp\left(-\frac{Rp}{c_3}\right). \tag{2.2.15}$$

Taking into account the inverse Laplace-transforms[+)]

$$\mathcal{L}^{-1}\left(e^{-\frac{Rp}{c_3}}\right) = \delta\left(\frac{R}{c_3} - t\right) ,$$

$$\mathcal{L}^{-1}\left(\bar{F}(\underline{x}, \underline{\xi}, p)\right) = \frac{R\nu}{c_3} \frac{\mathfrak{J}_1\left(\nu\sqrt{t^2 - R^2/c_3^2}\right)}{\sqrt{t^2 - R^2/c_3^2}} H\left(t - \frac{R}{c_3}\right) = F(R, t) ,$$

where

$$H\left(t - \frac{R}{c_3}\right) = \begin{cases} 0 & \text{for} \quad t < R/c_3 , \\ 1 & \text{for} \quad t > R/c_3 , \end{cases}$$

is the Heaviside function, and using the convolution formula

$$\mathcal{L}^{-1}(\bar{\sigma}, \bar{f}) = \int_0^t \sigma(\underline{x}, t-\tau) f(\underline{x}, \tau)\, d\tau = \int_0^t \sigma(\underline{x}, \tau) f(\underline{x}, t-\tau)\, d\tau ,$$

the function $\Sigma(\underline{\xi}, t)$ is represented in the form

$$\Sigma(\underline{\xi}, t) = \frac{1}{4\pi c_3^2} \left\{ \int_{V'} \frac{\sigma(\underline{x}, t - R/c_3)}{R(\underline{x}, \underline{\xi})} dV(\underline{x}) + \int_{V'} \frac{dV(\underline{x})}{R(\underline{x}, \underline{\xi})} \int_0^t \sigma(\underline{x}, t-\tau) F(\underline{x}, \underline{\xi}, \tau) d\tau \right\} . \quad (2.2.16)$$

Function $\mathfrak{J}_1(z)$ appearing in formula (2.2.17) is the Bessel function of the first kind and first order.

In the first right-hand integral of (2.2.16) the argument $t - \frac{R}{c_3}$ of the function σ appears, denoting the instant prior to instant t at which the integral is calculated. The time interval R/c_3 denotes the time required for the wave to travel from point $\underline{x}$ to point $\underline{\xi}$, where the function Σ is calculated.

+) Erdélyi (Editor): "Tables of integral transforms", Vol.II, Mc.Graw Hill, New York, 1954

Hence, the first integral represents the retarded potential. In particular, when $\sigma = f(t)\,\delta(\underline{x})$, i.e. when the disturbance is concentrated at the origin of the coordinate system and varies from instant $t = 0$ according to the function $f(t)$, then

$$(2.2.17)\quad \Sigma(\underline{x},t) = \frac{1}{4\pi c_3^2 R_0} f\left(t - \frac{R}{c_3}\right) + \frac{1}{4\pi c_3^2 R} \int_0^t f(t-\tau) F(R_0,\tau)\, d\tau ,$$

where

$$R_0 = \left(x_1^2 + x_2^2 + x_3^2\right)^{1/2} .$$

Returning back to the longitudinal wave the first part of (2.2.5) let us write it in the form

$$(2.2.18)\qquad \left(\nabla^2 - \frac{1}{c_1^2}\partial_t^2\right)\Phi(\underline{x},t) = -\frac{1}{c_1^2}\,\vartheta(\underline{x},t) .$$

The solution of this non-homogeneous equation (under homogeneous boundary conditions) is obtained directly from the solution (2.2.16), notations being suitably changed and assumed $\nu = 0$. Hence,

$$(2.2.19)\quad \vartheta(\underline{x},t) = \frac{1}{4\pi c_1^2} \int_V \frac{\vartheta(x, t - R/c_1)}{R(\underline{x},\underline{\xi})}\, dV(\underline{x}) .$$

The longitudinal wave is described by the retarded potential only.

Passing to a two-dimensional problem, let us make the assumption that potentials Σ and Φ are independent of x_3. Green's function of the two-dimensional problem should satisfy the equation (2.2.20)

$$\left(\partial_1^2+\partial_2^2-\frac{1}{c_3^2}\partial_t^2-\frac{\nu^2}{c_3^2}\right)G(x_1,x_2;\xi_1,\xi_2;t)=-2\pi\delta(x_1-\xi_1)\delta(x_2-\xi_1)\delta(t),$$

with homogeneous initial conditions and satisfaction of the condition $G\to 0$ for $\left|x_1^2+x_2^2\right|\to\infty$. Applying the Laplace-integral-transform to (2.2.20) and making use of the homogeneity of the initial conditions, we are led to

$$\left(\partial_1^2+\partial_2^2-\frac{1}{c_3^2}(p^2+\nu^2)\right)\bar{G}(x_1,x_2;\xi_1,\xi_2,p)=-2\pi\,\delta(x_1-\xi_1)\,\delta(x_2-\xi_2).$$

The solution to this problem has the form

$$\bar{G}(r,p)=K_0\left(r\sqrt{p^2+\nu^2}\right), \tag{2.2.21}$$

where

$$r=\left[(x_1-\xi_1)^2+(x_2-\xi_2)^2\right]^{1/2}.$$

Here $K_0(z)$ is the modified Bessel function of the third kind and zero order. Applying the inverse transform to (2.2.21) we obtain

$$G(r,t)=\frac{c_3}{\sqrt{(tc_3)^2-r^2}}\cos\left(\nu\sqrt{t^2-r^2/c_3^2}\right)H\left(t-\frac{r}{c_3}\right). \tag{2.2.22}$$

The knowledge of $\bar{G}$ enables us to determine the function $\bar{\Sigma}$ from the formula

$$\Sigma(\xi_1,\xi_2,p)=\frac{1}{2\pi c_3^2}\int_A\bar{\sigma}(x_1,x_2,p)\,\bar{G}(x_1,x_2;\xi_1,\xi_2,p)\,dx_1dx_2.$$

Applying the inverse Laplace-transform we are led to

(2.2.23)

$$\Sigma(\xi_1,\xi_2,t)=\frac{1}{2\pi c_3^2}\int_A dx_1 dx_2\int_0^t \frac{\sigma(x_1,x_2,t-\tau)\cos\left(\nu\sqrt{\tau^2-r^2/c_3^2}\right)}{\sqrt{(\tau c_3)^2-r^2}}H\left(\tau-\frac{r}{c_3}\right)d\tau .$$

In the particular case of a point disturbance $\sigma=\delta(x_1)\delta(x_2)f(t)$ varying in time from instant $t=0$ according to the function $f(t)$, Eq.(2.2.23) yields

$$\text{(2.2.23')}\quad \Sigma(x_1,x_2,t)=\frac{1}{2\pi c_3}\int_0^t \frac{f(t-\tau)}{\sqrt{(\tau c_3)^2-r_0^2}}\cos\left(\nu\sqrt{\tau^2-r_0^2/c_3^2}\right)H\left(\tau-\frac{r}{c_3}\right)d\tau$$

where $r_0=(x_1^2+x_2^2)^{1/2}$.

Passing to the longitudinal wave described by Eq.(2.2.7), we are able to present the solution to the two-dimensional problem in the form

$$\text{(2.2.24)}\quad \Phi(\xi_1,\xi_2,t)=\frac{1}{2\pi c_1}\int dx_1 dx_2\int \frac{\vartheta(x_1,x_2,t-\tau)}{\sqrt{(\tau c_1)^2-r^2}}H\left(\tau-\frac{r}{c_1}\right)d\tau .$$

Formula (2.2.23) has been used here, notations being suitably changed and assumption $\nu = 0$ made.

Let us consider the homogeneous equation of the rotational wave (2.2.6) under non-homogeneous boundary conditions

$$\text{(2.2.25)}\quad \Sigma(\underline{x},0)=g(\underline{x}),\qquad \dot{\Sigma}(\underline{x},0)=h(\underline{x}) .$$

Performing the Laplace-transform on the homogeneous equation (2.2.6) and taking into account the initial conditions (2.2.25), we are led to

$$\text{(2.2.26)}\quad \left(\nabla^2-\frac{1}{c_3^2}(p^2+\nu^2)\right)\Sigma(x,p)=-\frac{1}{c_3^2}\left(pg(\underline{x})+h(\underline{x})\right).$$

By combining Eqs.(2.2.10) and (2.2.26) accordingly and integrating over the infinite region, the following expression for the function Σ is obtained

$$\bar{\Sigma}(\underline{\xi}, p) = \frac{1}{4\pi c_3^2} \int_V \big(p g(\underline{x}) + h(\underline{x})\big) G(\underline{x}, \underline{\xi}, p)\, dV(\underline{x}), \quad (2.2.27)$$

or

$$\bar{\Sigma}(\underline{\xi}, p) = \frac{1}{4\pi c_3^2} \Big\{ \int_V (p g + h) \frac{\exp(-R/c_3)}{R(\underline{x}, \underline{\xi})}\, dV(\underline{x}) + \int_V \big(p g(\underline{x}) + h(\underline{x})\big) \bar{F}(\underline{x}, \underline{\xi}, p)\, dV(\underline{x}) \Big\} .$$

The inverse Laplace-transform leads to the formula

$$\Sigma(\underline{x}, t) = \frac{1}{4\pi c_3^2} \Big\{ \int_V \Big(h(\underline{\xi}) + g(\underline{\xi}) \frac{\partial}{\partial t} \Big) \frac{\delta\big(\frac{R}{c_3} - t\big)}{R(\underline{x}, \underline{\xi})}\, dV(\underline{\xi}) + \int_V \Big(h(\underline{\xi}) + g(\underline{\xi}) \frac{\partial}{\partial t} \Big) F(\underline{\xi}, \underline{x}, t)\, dV(\underline{\xi}) \Big\} . \quad (2.2.28)$$

Let us consider the first integral. Introducing the spherical coordinates (R, ϑ, ψ), the position coordinates of the sphere ξ_i can be expressed in terms of the coordinates of the center of the sphere x_i :

$$\xi_i = x_i + n_i R , \qquad i = 1, 2, 3$$

where

$$n_1 = \sin\vartheta \cos\psi , \qquad n_2 = \sin\vartheta \sin\psi , \qquad n_3 = \cos\vartheta$$

$$0 \leq \vartheta \leq \pi \qquad\qquad 0 \leq \psi \leq 2\pi .$$

By making use of the properties of Dirac's function

$$\int_V f(z)\,\delta(z-t)\,dz = f(t)$$

and observing that $dV = R^2 dR \sin\Theta\, d\Theta\, d\psi$, the first right-hand integral of Eq.(2.2.17) is transformed to

$$\frac{1}{4\pi c_3^2}\int_V h(\underline{\xi})\frac{\delta\left(\frac{R}{c_3}-t\right)}{R(\underline{x},\underline{\xi})}\,dV(\underline{\xi}) = \frac{t}{4\pi}\int_0^{2\pi} d\psi \int_0^{\pi} h(x_i+n_i c_3 t)\sin\Theta\, d\Theta\,.$$

The integral can be considered as the arithmetical mean of the function h over the surface of a sphere of radius $c_3 t$

Introducing the notation

$$M_{ct}\left\{h(x_i,t)\right\} = \frac{1}{4\pi}\int_0^{2\pi} d\psi \int_0^{\pi} h(x_i+n_i c_3 t)\sin\Theta\, d\Theta\,.$$

Eq.(2.2.28) is represented in the form

$$\Sigma(\underline{x},t) = t\,M_{ct}\left\{h(\underline{x},t)\right\} + \frac{\partial}{\partial t}\left[t\,M_{ct}\left\{g(\underline{x},t)\right\}\right] +$$

(2.2.29) $$+\frac{1}{4\pi c_3^2}\int_V \left[h(\underline{\xi}) + g(\underline{\xi})\frac{\partial}{\partial t}\right] F(\underline{\xi},\underline{x},t)\,dV(\underline{\xi})\,.$$

In a two-dimensional problem, following the analogous procedure and taking into account the function G from (2.2.22), the following expression for Σ is obtained.

$$\Sigma(x_1,x_2,t) = \frac{1}{2\pi c_3}\int_0^{c_3 t} r\,dr \int_0^{2\pi}\Big[h(x_1+r\cos\vartheta,\, x_2+r\sin\vartheta) +$$

$$+\, g(x_1+r\cos\vartheta,\, x_2+r\sin\vartheta)\frac{\partial}{\partial t}\Big]\;\frac{\cos\left(\nu\sqrt{t^2-r^2/c_3^2}\right)}{\sqrt{(c_3 t)^2-r^2}}\,d\vartheta\,,$$

(2.2.30)

where

$$r = \left[(x_1-\xi_1)^2+(x_2-\xi_2)^2\right]^{1/2}.$$

Solution of the one-dimensional problem (2.2.6) with non-homogeneous initial conditions is known+) and has the form

$$\Sigma(x_1,t) = \frac{1}{2}\left[g(x_1-c_3t)+g(x_1+c_3t)\right]+$$

$$-\frac{\nu t}{2c_3}\int_{x_1-c_3t}^{x_1+c_3t}\frac{J_1\left(\nu\sqrt{t^2-\left(\frac{x_1-\xi_1}{c_3}\right)^2}\right)}{\sqrt{t^2-\left(\frac{x_1-\xi_1}{c_3}\right)^2}}g(\xi_1)d\xi_1+$$

$$+\frac{1}{2c_3}\int_{x_1-c_3t}^{x_1+c_3t}J_0\left(\nu\sqrt{t^2-\left(\frac{x_1-\xi_1}{c_3}\right)^2}\right)h(\xi_1)d\xi_1 \quad . \qquad (2.2.31)$$

Let us return to Eq.(2.2.7) of the longitudinal waves. In absence of a source ($\vartheta = 0$) and with non-homogeneous initial conditions, the solution of (2.2.7)

$$\Phi(\underline{x},0)=k(\underline{x}), \qquad \dot{\Phi}(\underline{x},0)=f(\underline{x}), \qquad (2.2.32)$$

follows directly from formulae (2.2.29),(2.2.30) and (2.2.31) in which notations are changed and assumption $\nu = 0$ made. In such a way for the three-dimensional problem, one obtains

$$\Phi(\underline{x},t)=t\,M_{ct}\left\{f(x_i,t)\right\}+\frac{\partial}{\partial t}\left[t\,M_{ct}\left\{k(x_i,t)\right\}\right], \qquad (2.2.33)$$

+) P.M. Morse, H. Feshbach: "Methods of Theoretical Physics", Vol. 1 Mc. Graw-Hill, New York (1953), p.805.

where

$$M_{ct}\{f(x_i,t)\} = \frac{1}{4\pi}\int_0^{2\pi} d\psi \int_0^{\pi} f(x_i + n_i c_1 t)\sin\vartheta\, d\vartheta \ .$$

The well-know elastokinetics solution in the form of the "so-called" Poisson integrals is obtained.

For a two-dimensional problem the following formula is found

$$\Phi(x_1,x_2,t) = \frac{1}{2\pi c_1}\int_0^{c_1 t}\int_0^{2\pi}\Big[f(x_1 + r\cos\vartheta,\, x_2 + r\sin\vartheta) +$$

(2.2.34) $$+\, g(x_1 + r\cos\vartheta,\, x_2 + r\sin\vartheta)\frac{\partial}{\partial t}\Big]\frac{r\,dr\,d\vartheta}{\sqrt{(c_1 t)^2 - r^2}} \ .$$

Integration is performed over the circle $C_{c_1 t}$ of radius $c_1 t$

Finally, in a one-dimensional case function Φ is expressed by the well-known d'Alambert formula

(2.2.35) $$\Phi(x_1,t) = \frac{1}{2}\left[k(x_1 - c_1 t) + k(x_1 + c_1 t)\right] + \frac{1}{2c_1}\int_{x_1 - c_1 t}^{x_1 + c_1 t} f(\xi_1)\, d\xi_1 \ .$$

In the elastokinetics problem of longitudinal waves propagation, an important role is played by Kirchhoff's formula which makes it possible to express the value of a function $\Phi(\underline{\xi}, t)$ at the point $\underline{\xi} \in V$ in terms of surface integrals involving functions $\Phi(\underline{x}, t)$ and $\frac{\partial \Phi(\underline{x},t)}{\partial n}$ prescribed on $A, \underline{x} \in A$. This formula has the form (2.2.36)+)

+) B.B. Baker, E.T. Copson: "The mathematical theory of Huygen's principle", Oxford, Clarendon Press (1953), p. 37

$$\Phi(\underline{\xi}, t) = -\frac{1}{4\pi}\int_A \left\{ [\Phi]\frac{\partial}{\partial n}\left(\frac{1}{R}\right) - \frac{1}{c_1 R}\frac{\partial R}{\partial n}\left[\frac{\partial \Phi}{\partial t}\right] - \frac{1}{R}\left[\frac{\partial \Phi}{\partial n}\right]\right\} dA(\underline{x}) \, . \quad (2.2.36)$$

Here $\frac{\partial}{\partial n}$ denotes differentiation with respect to the outer normal to A. Expressions in brackets denote the retarded values $[\Phi] = \Phi\left(\underline{x}, t - \frac{R}{c_1}\right)$ etc. The formula holds true if $\underline{\xi} \in V$. If the point $\underline{\xi}$ is located outside the considered region, the value of the integral in (2.2.36) is zero. To derive this formula it is assumed that the first and second derivatives of the function are continuous inside the region V and on the surface A.

Let us prove the following formula for rotational waves

$$\Sigma(\underline{\xi}, t) = -\frac{1}{4\pi}\int_A \left\{ [\Sigma]\frac{\partial}{\partial n}\left(\frac{1}{R}\right) - \frac{1}{c_3 R}\frac{\partial R}{\partial n}\left[\frac{\partial \Sigma}{\partial t}\right] - \frac{1}{R}\left[\frac{\partial \Sigma}{\partial n}\right]\right\} dA(\underline{x}) +$$

$$-\frac{1}{4\pi}\int_A dA(\underline{x})\int_0^t \left\{ \left(F(R,t)\frac{\partial}{\partial n}\left(\frac{1}{R}\right) + \frac{1}{R}\frac{\partial R}{\partial n}\frac{\partial F}{\partial R}\right)\Sigma(\underline{x}, t-\tau) + \right.$$

$$\left. -\frac{1}{R}F(R,\tau)\frac{\partial \Sigma(\underline{x}, t-\tau)}{\partial n}\right\} d\tau \, , \qquad \underline{\xi} \in V \, , \quad (2.2.37)$$

$\Sigma(\underline{\xi}, t) = 0$, $\underline{\xi} \in C - V$, where C is the entire space.

The formula will be derived from equation (2.2.13) with the assumption $\overline{\mathfrak{S}} = 0$. With the aid of (2.2.11) and (2.2.15), simple transformations of Eq.(2.2.13) yield the following Eq.(2.2.38)

(2.2.38):

$$\bar{\Sigma}(\underline{\xi}, p) = -\frac{1}{4\pi}\int_A \Big\{ \bar{\Sigma} \frac{\partial}{\partial n}\Big(\frac{1}{R}\Big) \exp\Big(-\frac{R}{c_3}\sqrt{p^2+\nu^2}\Big) -$$

$$-\frac{1}{Rc_3}\frac{\partial R}{\partial n}\sqrt{p^2+\nu^2}\cdot\bar{\Sigma}\cdot\exp\Big(-\frac{R}{c_3}\sqrt{p^2+\nu^2}\Big) - \frac{\partial\bar{\Sigma}}{\partial n}\exp\Big(-\frac{R}{c_3}\sqrt{p^2+\nu^2}\Big)\Big\} dA(\underline{x}).$$

By introducing the notations

(2.2.39) $$[\bar{\Sigma}] = \bar{\Sigma}\cdot e^{-\frac{Rp}{c_3}}, \qquad \Big[\frac{\partial\bar{\Sigma}}{\partial n}\Big] = e^{-\frac{Rp}{c_3}}\frac{\partial\bar{\Sigma}}{\partial n},$$

Eq. (2.2.38) is reduced to

(2.2.40):

$$\bar{\Sigma}(\underline{\xi}, p) = -\frac{1}{4\pi}\int_A \Big\{ [\bar{\Sigma}] \frac{\partial}{\partial n}\Big(\frac{1}{R}\Big) - \frac{1}{c_3 R}\frac{\partial R}{\partial n}[p\,\bar{\Sigma}] - \frac{1}{R}\Big[\frac{\partial\bar{\Sigma}}{\partial n}\Big]\Big\} dA(\underline{x}) +$$

$$-\frac{1}{4\pi}\int_A \Big\{ F\bar{\Sigma}\cdot\frac{\partial}{\partial n}\Big(\frac{1}{R}\Big) + \frac{1}{R}\frac{\partial R}{\partial n}\Big(\bar{\Sigma}\frac{\partial\bar{F}}{\partial n}\Big) - \frac{1}{R}\Big(F\frac{\partial\bar{\Sigma}}{\partial n}\Big)\Big\} dA(\underline{x}).$$

Applying the inverse Laplace-transform to (2.2.40) and in compliance with

$$\mathcal{L}^{-1}[\bar{\Sigma}] = \int_0^t \Sigma(\underline{x}, t-\tau)\,\delta\Big(\frac{R}{c_3}-\tau\Big)d\tau = \Sigma\Big(\underline{x}, t-\frac{R}{c_3}\Big) = [\Sigma(\underline{x}, t)],$$

$$\mathcal{L}^{-1}[p\bar{\Sigma}] = \int_0^t \frac{\partial\Sigma(\underline{x}, t-\tau)}{\partial\tau}\,\delta\Big(\frac{R}{c_3}-\tau\Big)d\tau = \frac{\partial\Sigma(\underline{x}, t-R/c_3)}{\partial t} = \Big[\frac{\partial\Sigma}{\partial t}\Big],$$

$$\mathcal{L}^{-1}\Big[\frac{\partial\bar{\Sigma}}{\partial n}\Big] = \int_0^t \frac{\partial\Sigma(\underline{x}, t-\tau)}{\partial n}\,\delta\Big(\frac{R}{c_3}-\tau\Big)d\tau = \frac{\partial\Sigma(\underline{x}, t-R/c_3)}{\partial n} = \Big[\frac{\partial\Sigma}{\partial n}\Big],$$

formula (2.2.47) is obtained. This formula can be treated as a generalization of Kirchhoff's formula (constituting the integral formulation of Huygen's principle) to the Klein-Gordon wave equation.

Let us pass to the particular case of monochromatic vibrations. By inserting the expressions

$$\Sigma(\underline{x},t)=\Sigma^*(\underline{x})\,e^{-i\omega t}\,,\quad G(\underline{x},\underline{\xi},t)=G^*(\underline{x},\underline{\xi})\,e^{-i\omega t}$$

into (2.2.37) we are led to

$$\Sigma^*(\underline{\xi})=\frac{1}{4\pi}\int_A\left(\frac{\partial\Sigma^*}{\partial n}\frac{e^{ikR}}{R}-\Sigma^*\frac{\partial}{\partial n}\left(\frac{e^{ikR}}{R}\right)\right)dA(\underline{x}),\qquad \underline{\xi}\in V, \tag{2.2.41}$$

where

$$k=\frac{1}{c_3}\left(\omega^2-\nu^2\right)^{1/2},\qquad\qquad \nu^2=\frac{4\alpha}{J}\,.$$

This formula is true for $\omega>\nu$, since only then we are dealing with real values of the phase velocity. The presence of the frequency ω in the expression for the phase velocity $c=\frac{\omega}{k}$ indicates that the wave is dispersive.

The Helmholtz formula similar to (2.2.41) is obtained in the case of longitudinal waves

$$\Phi(\underline{\xi},t)=\frac{e^{-i\omega t}}{4\pi}\int_A\left(\frac{\partial\Phi^*}{\partial n}\frac{e^{i\sigma R}}{R}-\Phi^*\frac{\partial}{\partial n}\left(\frac{e^{i\sigma R}}{R}\right)\right)dA(\underline{x}),\ \sigma=\frac{\omega}{c_1},\ \underline{\xi}\in V\,. \tag{2.2.42}$$

The longitudinal wave is non-dispersive, since the phase velocity $c=c_1=\frac{\omega}{\sigma}$ is constant.

Moreover, let us shortly discuss the character of propagation of waves Σ and Φ. The insight into the mechanism is enabled owing to Poisson's integrals (2.2.33) and Eq.

(2.2.29) in the three-dimensional case, and Eqs.(2.2.34),(2.2.30) - the two-dimensional case. Let us begin from the three-dimensional case. Assume the disturbances $k(\underline{x}), f(\underline{x})$ (associated with longitudinal waves) to be contained in a certain region D_0 bounded by surface A. Select a point $\underline{x}$ outside the region D_0 and describe a sphere of radius $d = c_1 t$ around this point M (Fig. 2.2.2). If $t < \frac{d_1}{c_1}$, where d_1 is the shortest distance from the point $\underline{x}$ to the surface A_0, then the entire sphere M_{ct} will remain outside of D_0. Both functions k and f are zero on M_{ct}. Eq.(2.2.33) gives $\Phi(\underline{x},t) = 0$, the point $\underline{x}$ remains at rest. At instant $t_1 = \frac{d_1}{c_1}$ the surface of the sphere will touch the region D_0,

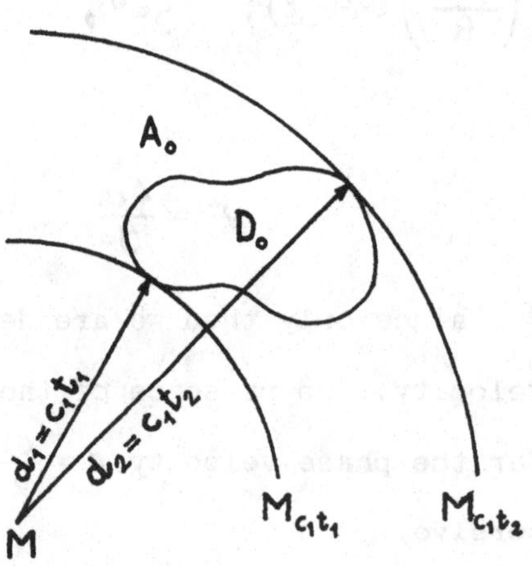

Fig. 2.2.2

the wavefront passing through the point $\underline{x}$. If $t > \frac{d_2}{c_1}$, where d_2 is the largest distance from the point $\underline{x}$ to the surface A_0, then the sphere M_{ct} will remain outside of D_0, and Eq.(2.2.33) will imply $\Phi(\underline{x},t) = 0$. The instant $t_2 = \frac{d_2}{c_1}$ corresponds to the passage of the wave end through the point $\underline{x}$, and at the next moment the function Φ assumes the zero value at $\underline{x}$. The mechanism of propagation of rotational waves is different. True enough, the two first terms of (2.2.29) represent the Poisson integrals of structure analogous, to that of the case of Φ , but a third term in the form of a surface integral remains. This integral does not vanish with the wave front passing through point $\underline{x}$.

Let us pass to the case of propagation of the two-dimensional wave described by Eq.(2.2.34). Assume the initial disturbance of the longitudinal wave to be limited in the plane Ox_1x_2 to a region D_0 bounded by the contour l . Functions k, f are different from zero within the region, but vanish outside. By selecting a point (x_1, x_2) outside the region D_0 and observing the propagation of the longitudinal wave provoked by the disturbances k, f in the domain D_0 .

Fig.(2.2.3) it can be found that at the instant $t < \frac{d_1}{c_1}$, d_1 being the shortest distance from the point $\underline{x} = (x_1, x_2, 0)$ to the contour l , the circle C_{ct} has no common points with D_0 . Functions k, f are zero over the entire circle $C_{c_1t_1}$ and formula (2.2.34) yields $\Phi(x_1, x_2, t) = 0$.

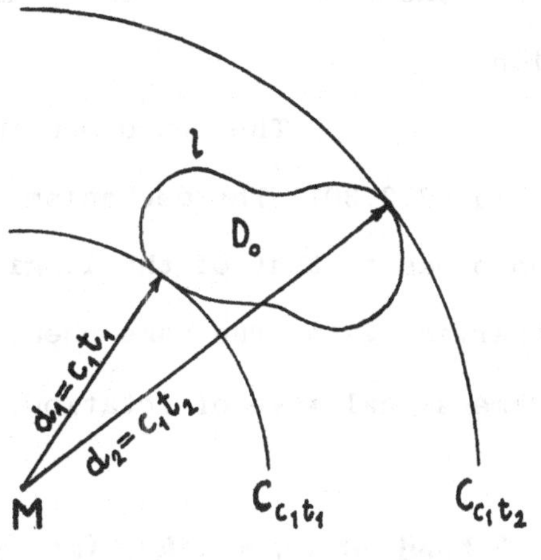

Fig. 2.2.3

At the instant $t_1 = \frac{d_1}{c_1}$ the wave front reaches the point $\underline{x}$. When $t > \frac{d_2}{c_1}$, d_2 denoting the largest distance from the point $\underline{x}$ to the points of the contour l , the circle C_{ct} will contain the entire region D_0 . The integration in (2.2.34) has to be performed over the region D_0 , since functions k , f are zero outside of this region. After the passage of the wave end through $\underline{x}$ at the instant $t_2 = \frac{d_2}{c_1}$, function Φ does not vanish, contrary to the three-dimensional case, which establishes the essential feature distinguishing the two- and three-dimensional cases. It may be observed, however, that with increasing time t , the wave function Φ decreases, owing to the factor $(c_1 t)^2$ appearing in the denominator of (2.2.34). The phenomenon of non-vanishing of the function Φ , after the wave passes, the point $\underline{x}$, is called the wave diffusion.

The two-dimensional wave of rotation is described by Eq.(2.2.30). The mechanism of propagation of the wave is analogous to that of the longitudinal wave; wave diffusion is observed again. The same phenomenon occurs in the case of a one-dimensional wave of rotation.

2. 3 Fundamental Solutions for an Infinite Elastic Space.*)

Let us consider the action of body forces and body moments in an infinite, micropolar, elastic space. Assume

*) Nowacki, W.: "Green functions for micropolar elasticity", Proc. Vibr. Probl. 10, 1, (1969), 3.

that the body forces and moments vary harmonically in time. Denoting the amplitudes of causes and effects by asterisks we obtain the following system of differential equations

$$(\mu+\alpha)\nabla^2\underline{u}^* + (\lambda+\mu-\alpha)\,\mathrm{grad\,div}\,\underline{u}^* + 2\alpha\,\mathrm{rot}\,\underline{\varphi}^* + \underline{X}^* + \rho\omega^2\underline{u}^* = 0, \tag{2.3.1}$$

$$(\gamma+\varepsilon)\nabla^2\underline{\varphi}^* + (\beta+\gamma-\varepsilon)\,\mathrm{grad\,div}\,\underline{\varphi}^* + 2\alpha\,\mathrm{rot}\,\underline{u}^* - 4\alpha\underline{\varphi}^* + \underline{Y}^* + J\omega^2\underline{\varphi}^* = 0. \tag{2.3.2}$$

The singular solutions of Eqs.(2.3.1) and (2.3.2), dependent on the distance R and the points $\underline{x}$ and $\underline{\xi}$, will be called the fundamental solutions. These solutions correspond to the displacements and rotations produced by the action of concentrated forces and concentrated moments.

The simplest way to obtain the solutions of the system of Eqs.(2.3.1) and (2.3.2) is to make use of the wave equations (1.8.7) (1.8.10). In the case of monochromatic vibrations these equations take the following form

$$(\nabla^2 + \sigma_1^2)\Phi^* = -\frac{1}{c_1^2}\vartheta^*, \tag{2.3.3}$$

$$(\nabla^2 + k_3^2)\Sigma^* = -\frac{1}{c_3^2}\sigma^*, \tag{2.3.4}$$

$$(\nabla^2 + k_1^2)(\nabla^2 + k_2^2)\underline{\Psi}^* = \frac{s}{c_4^2}\mathrm{rot}\,\underline{\eta}^* - \frac{1}{c_2^2}D_2\underline{\chi}^*, \tag{2.3.5}$$

$$(\nabla^2 + k_1^2)(\nabla^2 + k_2^2)\underline{H}^* = \frac{p}{c_2^2}\mathrm{rot}\,\underline{\chi}^* - \frac{1}{c_4^2}D_1\underline{\eta}^*. \tag{2.3.6}$$

Here we have introduced the following notation

$$\sigma_1 = \frac{\omega}{c_1}, \quad \sigma_2 = \frac{\omega}{c_2}, \quad k_3 = \frac{\omega}{c_3}\left(1 - \frac{\omega_0^2}{\omega^2}\right)^{1/2}, \quad \sigma_4 = \frac{\omega}{c_4}, \quad \omega_0^2 = \frac{4\alpha}{J},$$

$$c_1 = \left(\frac{\lambda + 2\mu}{\rho}\right)^{1/2}, \quad c_2 = \left(\frac{\mu + \alpha}{\rho}\right)^{1/2}, \quad c_3 = \left(\frac{\beta + 2\gamma}{J}\right)^{1/2}, \quad c_4 = \left(\frac{\gamma + \varepsilon}{J}\right)^{1/2},$$

$$s = \frac{2\alpha}{\mu + \alpha}, \quad p = \frac{2\alpha}{\gamma + \varepsilon}, \quad D_1 = \nabla^2 + \sigma_2^2, \quad D_2 = \nabla^2 + \sigma_4^2 - 2p.$$

The quantities k_1^2, k_2^2 constitute the roots of the equation

$$k^4 - k^2\left(\sigma_2^2 + \sigma_4^2 + p(s-2)\right) + \sigma_2^2\left(\sigma_4^2 - 2p\right) = 0. \tag{2.3.7}$$

We already encountered this equation at point 2.1 while discussing the plane waves. The following relations connect the amplitudes of the displacements and rotations and the amplitudes of the potentials and sources:

$$\begin{aligned} \underline{u}^* &= \operatorname{grad}\Phi^* + \operatorname{rot}\underline{\Psi}^*, & \operatorname{div}\underline{\Psi}^* &= 0, \\ \underline{\varphi}^* &= \operatorname{grad}\Sigma^* + \operatorname{rot}\underline{H}^*, & \operatorname{div}\underline{H}^* &= 0, \end{aligned} \tag{2.3.8}$$

and

$$\begin{aligned} \underline{X}^* &= \rho\left(\operatorname{grad}\vartheta^* + \operatorname{rot}\underline{\chi}^*\right), & \operatorname{div}\underline{\chi}^* &= 0, \\ \underline{Y}^* &= J\left(\operatorname{grad}\sigma^* + \operatorname{rot}\underline{\eta}^*\right), & \operatorname{div}\underline{\eta}^* &= 0. \end{aligned} \tag{2.3.9}$$

Let us, in turn, consider the homogeneous equations (2.3.3) - (2.3.6). The equation of longitudinal wave (2.3.3) is identical with that of longitudinal wave in the classical Hooke's medium. The singular solution satisfying this equation and the Sommerfeld conditions of radiation take the form $\Phi^* = A\frac{e^{i\sigma_1 R}}{R}$. The singular solution of Eq.(2.3.4) is given by the functions $\frac{e^{\pm i k_3 R}}{R}$. However, a physical meaning has only the solution $\frac{e^{i k_3 R}}{R}$ since only the expression

$$\mathrm{Re}\left[\frac{e^{-i\omega t} e^{i k_3 R}}{R}\right] = \frac{1}{R}\cos\omega\left(t - \frac{R}{v_3}\right), \qquad v_3 = c_3\left(1 - \frac{\omega_0^2}{\omega^2}\right)^{-1/2}$$

represents a divergent wave, moving out the place of disturbance to infinity. Since the phase velocity has to be real, we get $\omega^2 > \omega_0^2 = \frac{4\alpha}{J}$. Next, let us consider the homogeneous equation (2.3.5). The solution to this equation, according to T. Boggio's *) theorem, can be represented as the sum of two partial solutions

$$\underline{\Psi}^* = \underline{\Psi}'^* + \underline{\Psi}''^*,$$

satisfying Helmholtz vector equations

$$(\nabla^2 + k_1^2)\underline{\Psi}'^* = 0, \qquad (\nabla^2 + k_2^2)\underline{\Psi}''^* = 0.$$

Particular integrals of these equations are the functions $R^{-1} e^{\pm i k_j R}$, $j = 1,2$, $i = \sqrt{-1}$. The solutions $R^{-1} e^{i k_j R}$ are, however, only having a physical sense, because the expressions

$$\mathrm{Re}\left[e^{-i\omega t} \frac{e^{i k_j R}}{R}\right] = \frac{1}{R}\cos\omega\left(t - \frac{R}{v_j}\right), \quad v_j = \frac{\omega}{R_j}, \; j = 1,2.$$

are the only to express a wave moving from the place of pertur-

*) Boggio, T. op. cit. p. 56.

bation to infinity. The solution of the homogeneous Eq.(2.3.5) will, therefore, take the form:

$$\underline{\Psi}^* = \underline{A}\frac{e^{ik_1R}}{R} + \underline{B}\frac{e^{ik_2R}}{R}. \tag{2.3.10}$$

An analogous solution of the homogeneous Eq.(2.3.6) is presented by the function:

$$\underline{H}^* = \underline{C}\frac{e^{ik_1R}}{R} + \underline{D}\frac{e^{ik_2R}}{R}. \tag{2.3.11}$$

In the waves $\underline{\Psi}^*$, $\underline{H}^*$, real phase velocities are the only ones that can occur. We must, therefore, have $k_1^2 > 0$, $k_2^2 > 0$. The first condition is satisfied for a positive determinant of Eq.(2.3.7). The other condition will be satisfied if $\sigma_4^2 > 2p$ or if $\omega^2 > \frac{4\alpha}{J}$. This follows from the relation $k_1^2 k_2^2 = \sigma_2^2(\sigma_4^2 - 2p) > 0$. In the expressions (2.3.10), (2.3.11), there are two waves undergoing dispersion (because k_1 and k_2 are functions of the frequency ω). The micro-rotation wave Σ^* will exist if $\sigma_3^2 > 0$. This condition leads to the inequality $\omega^2 > \frac{4\alpha}{J}$.

Let us consider the action of body forces. Let us observe that the lack of body couples ($\underline{Y}$ = 0) results in $\sigma^* = 0$, $\eta^* = 0$. No micro-rotation wave will occur in the infinite elasticspace ($\Sigma^* = 0$). The following set of equations remains to be solved:

$$(\nabla^2 + \sigma_1^2)\Phi^* = -\frac{1}{c_1^2}\vartheta^*, \tag{2.3.12}$$

$$(\nabla^2 + k_1^2)(\nabla^2 + k_2^2)\underline{\Psi}^* = -\frac{1}{c_2^2} D_2 \underline{\chi}^*, \tag{2.3.13}$$

$$(\nabla^2 + k_1^2)(\nabla^2 + k_2^2)\underline{H}^* = \frac{p}{c_2^2} \operatorname{rot} \underline{\chi}^*. \tag{2.3.14}$$

If the forces $\underline{X}$ are distributed over a closed region B, the quantities $\underline{\vartheta}^*$ and $\underline{\chi}^*$ will be determined from the following equations+)

$$\vartheta^*(\underline{x}) = -\frac{1}{4\pi\rho} \int_B X_j^*(\underline{\xi}) \frac{\partial}{\partial x_j}\left(\frac{1}{R(\underline{\xi}, \underline{x})}\right) dV(\underline{\xi}) \tag{2.3.15}$$

$$\chi_i^*(\underline{x}) = -\frac{1}{4\pi\rho} \int_B \epsilon_{ijk} X_j^*(\underline{\xi}) \frac{\partial}{\partial x_k}\left(\frac{1}{R(\underline{\xi}, \underline{x})}\right) dV(\underline{\xi}), \quad i,j,k = 1,2,3. \tag{2.3.16}$$

By introducing in these equations the formula

$$X_j^* = \delta(x_1)\delta(x_2)\delta(x_3)\delta_{1j}, \qquad j = 1,2,3,$$

expressing a concentrated force at the origin in the direction of the x_1-axis, we obtain:

$$\vartheta^* = -\frac{1}{4\pi\rho} \frac{\partial}{\partial x_1}\left(\frac{1}{R}\right), \qquad \chi_1^* = 0,$$

$$\chi_2^* = -\frac{1}{4\pi\rho} \frac{\partial}{\partial x_3}\left(\frac{1}{R}\right), \quad \chi_3^* = -\frac{1}{4\pi\rho} \frac{\partial}{\partial x_2}\left(\frac{1}{R}\right) \tag{2.3.17}$$

$$R = (x_1^2 + x_2^2 + x_3^2)^{1/2}.$$

+) V.D. Kupradze: Dynamical Problems in Elasticity. Progress in Solid Mechanics. V.3, Amsterdam, 1963

It remains to solve the equations:

(2.3.18) $$(\nabla^2+\sigma_1^2)\Phi^* = \frac{1}{4\pi\rho c_1^2}\frac{\partial}{\partial x_1}\left(\frac{1}{R}\right),$$

(2.3.19) $$(\nabla^2+k_1^2)(\nabla^2+k_2^2)\Psi_2^* = -\frac{1}{4\pi\rho c_2^2}D_2\frac{\partial}{\partial x_3}\left(\frac{1}{R}\right),$$

(2.3.20) $$(\nabla^2+k_1^2)(\nabla^2+k_2^2)\Psi_3^* = \frac{1}{4\pi\rho c_2^2}D_2\frac{\partial}{\partial x_2}\left(\frac{1}{R}\right),$$

(2.3.21) $$(\nabla^2+k_1^2)(\nabla^2+k_2^2)H_1^* = -\frac{p}{4\pi\rho c_2^2}(\nabla^2-\sigma_1^2)\left(\frac{1}{R}\right),$$

(2.3.22) $$(\nabla^2+k_1^2)(\nabla^2+k_2^2)H_2^* = \frac{p}{4\pi\rho c_2^2}\frac{\partial^2}{\partial x_1\partial x_2}\left(\frac{1}{R}\right),$$

(2.3.23) $$(\nabla^2+k_1^2)(\nabla^2+k_2^2)H_3^* = \frac{p}{4\pi\rho c_2^2}\frac{\partial^2}{\partial x_1\partial x_3}\left(\frac{1}{R}\right).$$

The solution of Eq.(2.3.18) is known from classical elastokinetics (dynamic theory of elasticity):

(2.3.24) $$\Phi^*(\underline{x}) = -\frac{1}{4\pi\rho\omega^2}\frac{\partial}{\partial x_1}\left(\frac{e^{i\sigma_1 R}-1}{R}\right).$$

Equations (2.3.19),(2.3.20) will be solved by applying Fourier--integral-transformation of the exponential type. Thus, for instance, the solution of the equation for Ψ_2^* will be presented in the form of the integral:

(2.3.25) $$\Psi_2^*(\underline{x}) = \frac{1}{8\pi^3\rho c_2^2}\frac{\partial}{\partial x_3}\iiint_{-\infty}^{+\infty}\frac{(\underline{\alpha}^2-\sigma_4^2+2p)\,e^{-i\alpha_k x_k}\,d\alpha_1 d\alpha_2 d\alpha_3}{\underline{\alpha}^2(\underline{\alpha}^2-k_1^2)(\underline{\alpha}^2-k_2^2)},$$

where

$$\underline{\alpha}^2 = \alpha_1^2+\alpha_2^2+\alpha_3^2 .$$

Bearing in mind that

$$\iiint_{-\infty}^{+\infty} \frac{e^{-i\alpha_k x_k}\, d\alpha_1 d\alpha_2 d\alpha_3}{\underline{\alpha}^2 - k_j^2} = 2\pi^2 \frac{e^{ik_j R}}{R} ,$$

the function Ψ_2^* can be presented in the form:

$$\Psi_2^* = \frac{1}{4\pi\rho\omega^2} \frac{\partial}{\partial x_3}\left(A_1 \frac{e^{ik_1 R}}{R} + A_2 \frac{e^{ik_2 R}}{R} + A_3 \frac{1}{R} \right) , \qquad (2.3.26)$$

where

$$A_1 = \frac{\sigma_2^2 - k_2^2}{k_1^2 - k_2^2} , \qquad A_2 = \frac{\sigma_2^2 - k_1^2}{k_2^2 - k_1^2} , \qquad A_3 = -1 .$$

Solving Eq.(2.3.20) in an analogous manner, we have

$$\Psi_3^* = -\frac{1}{4\pi\rho\omega^2} \frac{\partial}{\partial x_2}\left(A_1 \frac{e^{ik_1 R}}{R} + A_2 \frac{e^{ik_2 R}}{R} + A_3 \frac{1}{R} \right) . \qquad (2.3.27)$$

Application of Fourier-integral-transformations to the set of Eqs.(2.3.21) - (2.3.23) yields

$$H_1^* = \frac{p}{4\pi\rho c_2^2}\left[\frac{e^{ik_1 R} - e^{ik_2 R}}{R(k_1^2 - k_2^2)} + \frac{\partial^2}{\partial x_1^2}\left(B_1 \frac{e^{ik_1 R}}{R} + B_2 \frac{e^{ik_2 R}}{R} + B_3 \frac{1}{R} \right) \right] , \qquad (2.3.28)$$

$$H_2^* = \frac{p}{4\pi\rho c_2^2} \frac{\partial^2}{\partial x_1 \partial x_2}\left(B_1 \frac{e^{ik_1 R}}{R} + B_2 \frac{e^{ik_2 R}}{R} + B_3 \frac{1}{R} \right) , \qquad (2.3.29)$$

$$H_3^* = \frac{p}{4\pi\rho c_2^2} \frac{\partial^2}{\partial x_1 \partial x_3}\left(B_1 \frac{e^{ik_1 R}}{R} + B_2 \frac{e^{ik_2 R}}{R} + B_3 \frac{1}{R} \right) , \qquad (2.3.30)$$

where

$$B_1 = \frac{1}{k_1^2 (k_1^2 - k_2^2)} , \qquad B_2 = \frac{1}{k_2^2 (k_2^2 - k_1^2)} , \qquad B_3 = \frac{1}{k_1^2 k_2^2} .$$

The displacements $\underline{u}$ and the rotations $\underline{\varphi}$ will be found from

(2.3.8).

Since $\Sigma^* = 0$, therefore:

$$u_1^* = \partial_1\Phi^* + \partial_2\Psi_3^* - \partial_3\Psi_2^*, \quad u_2^* = \partial_2\Phi^* - \partial_1\Psi_3^*, \quad u_3^* = \partial_3\Phi^* + \partial_1\Psi_2^*,$$

(2.3.31)

$$\varphi_1^* = \partial_2 H_3^* - \partial_3 H_2^*, \quad \varphi_2^* = \partial_3 H_1^* - \partial_1 H_3^*, \quad \varphi_3^* = \partial_1 H_2^* - \partial_2 H_1^*.$$

As a result, we shall obtain the following equations for the amplitudes $\underline{u}$ and φ :

$$u_j^* = U_j^{*(1)} = \frac{1}{4\pi\rho\omega^2}\left\{\left(A_1 k_1^2 \frac{e^{ik_1R}}{R} + A_2 k_2^2 \frac{e^{ik_2R}}{R}\right)\delta_{1j} + \right.$$

$$(2.3.32) \qquad \left. + \partial_1\partial_j\left(A_1\frac{e^{ik_1R}}{R} + A_2\frac{e^{ik_2R}}{R} + A_3\frac{e^{i\sigma_1R}}{R}\right)\right\},$$

$$(2.3.33) \qquad \varphi_j^* = \Phi_j^{*(1)} = \frac{p}{4\pi\rho c_2^2(k_1^2-k_2^2)}\epsilon_{1jk}\frac{\partial}{\partial x_k}\left(\frac{e^{ik_1R} - e^{ik_2R}}{R}\right),$$

$$j, k = 1, 2, 3.$$

We have obtained three components of the displacement vector $U_j^{(1)}$ and three components of the vector of rotation $\Phi_j^{(1)}$. We now displace the concentrated force from the origin to the point $\underline{\xi}$, and let it act parallel to the x_ℓ-axis. Then, Eqs.(2.3.32) and (2.3.33) become:

$$u_j^* = U_j^{*(\ell)} = \frac{1}{4\pi\rho\omega^2}\left\{\left(A_1 k_1^2 \frac{e^{ik_1R}}{R} + A_2 k_2^2 \frac{e^{ik_2R}}{R}\right)\delta_{j\ell} + \right.$$

$$(2.3.34) \qquad \left. + \partial_\ell\partial_j\left(A_1\frac{e^{ik_1R}}{R} + A_2\frac{e^{ik_2R}}{R} + A_3\frac{e^{i\sigma_1R}}{R}\right)\right\},$$

and

$$\varphi_j^* = \Phi_j^{*(\ell)} = \frac{p}{4\pi\rho c_2^2(k_1^2-k_2^2)}\,\epsilon_{\ell jk}\frac{\partial}{\partial x_k}\left(\frac{e^{ik_1R}-e^{ik_2R}}{R}\right), \quad j,k,\ell=1,2,3. \quad (2.3.35)$$

In Eqs.(2.3.34) and (2.3.35) R takes a different meaning. We have

$$R = \left[(x_1-\xi_1)^2+(x_2-\xi_2)^2+(x_3-\xi_3)^2\right]^{1/2}.$$

Thus, we have obtained the displacement tensor $U_j^{*(\ell)}(\underline{x},\underline{\xi})$ and the rotation tensor $\Phi_j^{*(\ell)}(\underline{x},\underline{\xi})$. These tensors constitute two symmetric matrices.

Let us introduce in Eqs.(2.3.34) and (2.3.35) $\alpha = 0$, thus, passing to classical elastokinetics, we have:

$$U_j^{*(\ell)} = \frac{e^{i\tau R}}{4\pi\mu R}\delta_{j\ell} - \frac{1}{4\pi\rho\omega^2}\frac{\partial^2}{\partial x_j\partial x_\ell}\left(\frac{e^{i\sigma R}-e^{i\tau R}}{R}\right), \quad (2.3.36)$$

$$\Phi_j^{*(\ell)} = 0 \quad , \quad j,\ell = 1,2,3$$

with the notations

$$\tau = \frac{\omega}{c_2^\circ}, \quad c_2^\circ = \left(\frac{\mu}{\rho}\right)^{1/2}, \quad \sigma = \frac{\omega}{c_1}, \quad c_1 = \left(\frac{\lambda+2\mu}{\rho}\right)^{1/2}.$$

Let us return to Eqs.(2.3.32) and (2.3.33) and observe, that the concentrated force parallel to x_1 does not produce rotation φ_1^*. We have $\varphi_1^* = \Phi_1^{*(1)} = 0$. This results in the fact that the components $\varkappa_{1j}$ (j = 1,2,3) of the curvature-twist tensor are zero. The components of the strain tensor γ_{ji} are different to zero.

Equations (2.3.34) and (2.3.35) express waves of three types. Waves connected with the values k_1, k_2 undergo dispersion.

Let us consider the action of body couples. Since $\underline{X} = 0$, therefore also $\vartheta = 0$, $\underline{\chi} = 0$. No longitudinal wave ($\Phi_j^* = 0$) will occur in the infinite space. We must now solve the set of equations

$$(\nabla^2 + k_3^2)\Sigma^* = -\frac{1}{c_3^2}\sigma^*, \tag{2.3.37}$$

$$(\nabla^2 + k_1^2)(\nabla^2 + k_2^2)\underline{\Psi}^* = \frac{s}{c_4^2}\operatorname{rot}\underline{\eta}^*, \tag{2.3.38}$$

$$(\nabla^2 + k_1^2)(\nabla^2 + k_2^2)\underline{H}^* = -\frac{1}{c_4^2}D_1\underline{\eta}^*. \tag{2.3.39}$$

If the body couples $\underline{Y}$ are distributed over a closed region B, the quantities σ^* and $\underline{\eta}^*$ will be found from the equations:

$$\sigma^*(\underline{x}) = -\frac{1}{4\pi J}\int_B Y_j^*(\underline{\xi})\frac{\partial}{\partial x_j}\left(\frac{1}{R(\underline{\xi},\underline{x})}\right)dV(\underline{\xi}), \tag{2.3.40}$$

$$\eta_i^*(\underline{x}) = -\frac{1}{4\pi J}\int_B \epsilon_{ijk}Y_j^*(\underline{\xi})\frac{\partial}{\partial x_k}\left(\frac{1}{R(\underline{\xi},\underline{x})}\right)dV(\underline{\xi}), \tag{2.3.41}$$

$$i,j,k = 1,2,3.$$

On introducing in these equations the expression

$$Y_j^*(\underline{x}) = \delta(x_1)\delta(x_2)\delta(x_3)\delta_{1j}, \qquad j = 1,2,3$$

- that is, a concentrated body couple acting at the origin in the x_1- direction - we obtain:

$$\sigma^* = -\frac{1}{4\pi J}\frac{\partial}{\partial x_1}\left(\frac{1}{R}\right), \quad \eta_1^* = 0, \quad \eta_2^* = \frac{1}{4\pi J}\frac{\partial}{\partial x_3}\left(\frac{1}{R}\right), \tag{2.3.42a}$$

and

$$\eta_3^* = -\frac{1}{4\pi\mathcal{J}} \frac{\partial}{\partial x_2}\left(\frac{1}{R}\right). \tag{2.3.42b}$$

Upon solving Eqs.(2.3.37) to (2.3.39) in the same manner as was done for the body forces, we find:

$$\Sigma^* = -\frac{1}{4\pi c_3^2 k_3^2 \mathcal{J}} \frac{\partial}{\partial x_1}\left(\frac{e^{ik_3R}-1}{R}\right), \tag{2.3.43}$$

$$\Psi_j^* = \frac{s}{4\pi c_4^2 \mathcal{J}}\left\{\frac{e^{ik_1R}-e^{ik_2R}}{R(k_1^2-k_2^2)} + \partial_1\partial_j\left(B_1\frac{e^{ik_1R}}{R} + B_2\frac{e^{ik_2R}}{R} + \frac{B_3}{R}\right)\right\}, \tag{2.3.44}$$

$$H_j^* = \frac{1}{4\pi c_4^2 \mathcal{J}} \epsilon_{1jk} \frac{\partial}{\partial x_k}\left(C_1\frac{e^{ik_1R}}{R} + C_2\frac{e^{ik_2R}}{R} + C_3\frac{1}{R}\right), \tag{2.3.45}$$

where

$$C_1 = \frac{k_1^2-\sigma_2^2}{k_1^2(k_1^2-k_2^2)}, \qquad C_2 = \frac{k_2^2-\sigma_2^2}{k_2^2(k_1^2-k_2^2)}, \qquad C_3 = -\frac{\sigma_2^2}{k_1^2 k_2^2}.$$

The displacements and rotations will be found from the equations

$$u_1^* = \partial_2\Psi_3^* - \partial_3\Psi_2^*, \quad u_2^* = \partial_3\Psi_1^* - \partial_1\Psi_3^*, \quad u_3^* = \partial_1\Psi_2^* - \partial_2\Psi_1^*$$

$$\varphi_1^* = \partial_1\Sigma^* + \partial_2 H_3^* - \partial_3 H_2^*, \quad \varphi_2^* = \partial_2\Sigma^* - \partial_1 H_3^*, \quad \varphi_3^* = \partial_3\Sigma^* + \partial_1 H_2^*. \tag{2.3.46}$$

Upon substituting (2.3.43) - (2.3.45) in (2.3.46), we obtain:

$$u_j^* = V_j^{*(1)} = \frac{s\,\epsilon_{1jk}}{4\pi\mathcal{J}c_4^2(k_1^2-k_2^2)} \frac{\partial}{\partial x_k}\left(\frac{e^{ik_1R}-e^{ik_2R}}{R}\right), \tag{2.3.47}$$

$$\varphi_j^* = W_j^{*(1)} = \frac{1}{4\pi\mathcal{J}c_4^2}\left\{\left(k_1^2 C_1\frac{e^{ik_1R}}{R} + k_2^2 C_2\frac{e^{ik_2R}}{R}\right)\delta_{1j} + \right.$$

$$\left. + \partial_1\partial_j\left(C_1\frac{e^{ik_1R}}{R} + C_2\frac{e^{ik_2R}}{R} + C_3\frac{e^{ik_3R}}{R}\right)\right\}, \qquad k,j=1,2,3. \tag{2.3.48}$$

Upon moving the concentrated body couple to the point $\underline{\xi}$, and directing the body couple vector parallel to the x_ℓ-axis, we obtain the Green tensor of displacement $V_j^{*(\ell)}(\underline{x},\underline{\xi})$ and the rotation tensor $W_j^{*(\ell)}(\underline{x},\underline{\xi})$.

Thus, for example, we have

$$(2.3.49)\quad V_j^{*(\ell)}(\underline{x},\underline{\xi}) = \frac{s}{4\pi \mathcal{J} c_4^2 (k_1^2 - k_2^2)} \epsilon_{\ell jk} \frac{\partial}{\partial x_k}\left(\frac{e^{ik_1 R} - e^{ik_2 R}}{R}\right), \quad \ell, j, k = 1,2,3 ,$$

where

$$R = \left[(x_1 - \xi_1)^2 + (x_2 - \xi_2)^2 + (x_3 - \xi_3)^2\right]^{1/2}.$$

Upon returning to Eqs.(2.3.47) - (2.3.48), let us observe that the action of the concentrated body couple $Y_j^* = \delta(x_1)\delta(x_2)\delta(x_3)\delta_{1j}$ produces zero displacement in the direction of the axis x_1 $(V_1^{*(1)} = 0)$, therefore, also, $\gamma_{11} = 0$. Since k_1, k_2, k_3 are functions of the frequency ω , all the types of waves occurring in the expressions (2.3.47) and (2.3.48) undergo dispersion.

Let a body force $X_j = \delta(x_1)\delta(x_2)\delta_{1j}\, e^{i\omega t}$ act in the infinite elastic body in the direction of the x_1-axis and let these forces be uniformly distributed along the x_3-axis. In this case, the displacements and the rotations are independent of the variable x_3, and we are concerned with the two-dimensional problem.

The relevant equations for the two-dimensional problem will be obtained from the equations of the foregoing sections by means of the principle of superposition.

Let us begin from the Green function for displace

ment $U_j^{*(1)}$, Eq.(2.3.32), assuming that the concentrated force acts at the point $(0,0,\xi_3)$ in the direction of the x_1-axis. On integrating the function $U_j^{*(1)}$ along the x_3-axis from $-\infty$ to $+\infty$, we shall find the corresponding equations for the displacement in the two-dimensional problem.

Let us observe that

$$\int_{-\infty}^{\infty} \frac{\exp(ik_\alpha\sqrt{r^2+\xi_3^2})}{\sqrt{r^2+\xi_3^2}}\,d\xi_3 = 2K_0(-ik_\alpha r), \quad \alpha = 1,2, \tag{2.3.50}$$

$$r = (x_1^2 + x_2^2)^{1/2},$$

where $K_0(x)$ is the modified Bessel function of the third kind. On the other hand, we have

$$2K_0(-ik_\alpha r) = \pi i H_0^{(1)}(k_\alpha r), \quad \alpha = 1,2, \tag{2.3.51}$$

where $H_0^{(1)}(k_\alpha r)$ is a Hankel function.

If now we integrate the displacements $U_j^{*(1)}$ of (2.3.32) along the x_3-axis, we obtain, bearing in mind (2.3.50) and (2.3.51):

$$U_j^{*(1)}(x_1,x_2;0,0) = \frac{i}{2\rho\omega^2}\Big[\big(A_1k_1^2H_0^{(1)}(k_1r)+A_2k_2^2H_0^{(1)}(k_2r)\big)\delta_{1j} +$$

$$+\,\partial_1\partial_j\big(A_1H_0^{(1)}(k_1r)+A_2H_2^{(1)}(k_2r)+A_3H_0^{(1)}(\sigma_1r)\big)\Big], \quad j=1,2. \tag{2.3.52}$$

In a similar manner, we shall also determine the rotation $\Phi_i^{*(1)}(x_1, x_2; 0,0)$. In Eq.(2.3.33), we obtain:

$$\Phi_1^{*(1)} = 0, \qquad \Phi_2^{*(1)} = 0, \tag{2.3.53a}$$

(2.3.53b) $$\Phi_3^{*(1)} = -\frac{p\epsilon_{13k}}{2\varrho c_2^2(k_1^2 - k_2^2)}\frac{\partial}{\partial x_2}\left[H_0^{(1)}(k_1 r) - H_0^{(1)}(k_2 r)\right].$$

We must now direct the linear force parallel to x_ℓ (ℓ = 1,2), and move it to the point $\underline{\xi} = (\xi_1, \xi_2)$. Then

$$U_j^{*(\ell)}(x_1, x_2; \xi_1, \xi_2) = \frac{i}{2\varrho\omega^2}\Big[\big(A_1 k_1^2 H_0^{(1)}(k_1 r) + A_2 k_2^2 H_0^{(1)}(k_2 r)\big)\partial_{j\ell} +$$

(2.3.54) $$+\partial_\ell\partial_j\big(A_1 H_0^{(1)}(k_1 r) + A_2 H_0^{(1)}(k_2 r) + A_3 H_0^{(1)}(\sigma_1 r)\big)\Big], \quad j, \ell = 1,2.$$

We shall now describe in brief the other method for finding the Green function. Use will be made of the stress functions $\underline{F}$ and $\underline{G}$ generalized by N. Sandru+) . These functions are connected with the displacements and rotations by the relations:

(2.3.55) $$\underline{u} = \Box_1\Box_4\underline{F} - \operatorname{grad}\operatorname{div}\Gamma\underline{F} - 2\alpha\operatorname{rot}\Box_3\underline{G},$$

(2.3.56) $$\underline{\varphi} = \Box_2\Box_3\underline{G} - \operatorname{grad}\operatorname{div}\Theta\underline{G} - 2\alpha\operatorname{rot}\Box_1\underline{F},$$

where

$$\Gamma = (\lambda + \mu - \alpha)\Box_4 - 4\alpha^2, \qquad \Theta = (\beta + \gamma - \varepsilon)\Box_2 - 4\alpha^2.$$

By introducing (2.3.55) and (2.3.56) in the set of Eqs.(2.3.1), (2.3.2), we obtain:

(2.3.57) $$\Box_1(\Box_2\Box_4 + 4\alpha^2\nabla^2)\underline{F} + \underline{X} = 0,$$

(2.3.58) $$\Box_3(\Box_2\Box_4 + 4\alpha^2\nabla^2)\underline{G} + \underline{Y} = 0.$$

+) N. Sandru, loc. cit. p.53.

The particular usefulness of these equations for the determination of the Green function is evident. It suffices to find a particular integral of these equations and to determine the displacement $\underline{u}$ and the rotation $\underline{\varphi}$ from Eqs. (2.3.55) and (2.3.56).

From Eqs.(2.3.57),(2.3.58), we find that $\underline{F} = 0$ for no body forces and $\underline{G} = 0$ for no body couples. By considering harmonic body forces and couples, Eqs.(2.3.57) and (2.3.58) can be reduced to the form:

$$(\nabla^2 + k_1^2)(\nabla^2 + k_2^2)(\nabla^2 + \sigma_1^2)\underline{F}^* + \varkappa \underline{X}^* = 0, \qquad (2.3.59)$$

$$(\nabla^2 + k_1^2)(\nabla^2 + k_2^2)(\nabla^2 + k_3^2)\underline{G}^* + \sigma \underline{Y}^* = 0, \qquad (2.3.60)$$

where

$$\varkappa = \frac{1}{(\lambda+2\mu)(\mu+\alpha)(\gamma+\varepsilon)}, \quad \sigma = \frac{1}{(\beta+2\gamma)(\mu+\alpha)(\gamma+\varepsilon)} .$$

Let us observe that the solution of the homogeneous Eqs.(2.3.59) and (2.3.60) has the form:

$$\underline{F}^* = \underline{A}\,\frac{e^{ik_1R}}{R} + \underline{B}\,\frac{e^{ik_2R}}{R} + \underline{C}\,\frac{e^{i\sigma_1R}}{R}, \qquad (2.3.61)$$

$$\underline{G}^* = \underline{D}\,\frac{e^{ik_1R}}{R} + \underline{E}\,\frac{e^{ik_2R}}{R} + \underline{H}\,\frac{e^{ik_3R}}{R}. \qquad (2.3.62)$$

It is seen that the first two wave terms of (2.3.61) undergo dispersion. In Eq.(2.3.62) all the three wave terms are dispersed. Let us quote the equations for the amplitude of displacement and rotation:

$$\underline{u}^* = (\lambda + 2\mu)(\gamma + \varepsilon)(\nabla^2 + k_3^2)(\nabla^2 + \sigma_4^2 - 2p)\underline{F}^* +$$

$$- (\gamma + \varepsilon)(\lambda + \mu - \alpha)\,\mathrm{grad\,div}\,[(\nabla^2 + \sigma_4^2 - 2p - \eta)]\underline{F}^* +$$

$$- 2\alpha(\beta + 2\gamma)(\nabla^2 + k_3^2)\,\mathrm{rot}\,\underline{G}^* , \tag{2.3.63}$$

$$\underline{\varphi}^* = (\mu + \alpha)(\beta + 2\gamma)(\nabla^2 + \sigma_2^2)(\nabla^2 + k_3^2)\underline{G}^* +$$

$$- (\beta + \gamma - \varepsilon)(\gamma + \varepsilon)\,\mathrm{grad\,div}\,[\nabla^2 + \sigma_2^2 - 2s - \zeta]\underline{G}^* +$$

$$- 2\alpha(\lambda + 2\mu)(\nabla^2 + \sigma_1^2)\,\mathrm{rot}\,\underline{F}^* , \tag{2.3.64}$$

where

$$\zeta = \frac{4\alpha^2}{(\beta + \gamma - \varepsilon)(\gamma + \varepsilon)} , \qquad \eta = \frac{4\alpha^2}{(\gamma + \varepsilon)(\lambda + \mu - \alpha)} .$$

Let us consider first the action of body forces. Since $\underline{Y}^* = 0$, therefore also $\underline{G}^* = 0$. It remains to consider Eq.(2.3.59) and to set $\underline{G}^* = 0$ in Eqs.(2.3.63),(2.3.64).

By applying to (2.3.59) the Fourier exponential transformation, and introducing the new notations

$$\mu_1 = k_1 , \qquad \mu_2 = k_2 , \qquad \mu_3 = \sigma_1 ,$$

we obtain

$$\underline{F}^* = \left(\frac{\underline{I}_1^*}{(\mu_1^2 - \mu_2^2)(\mu_1^2 - \mu_3^2)} + \frac{\underline{I}_2^*}{(\mu_2^2 - \mu_1^2)(\mu_2^2 - \mu_3^2)} + \frac{\underline{I}_3^*}{(\mu_3^2 - \mu_1^2)(\mu_3^2 - \mu_2^2)} \right) \tag{2.3.65}$$

The vector functions $\underline{I}_1^*$, $\underline{I}_2^*$, $\underline{I}_3^*$, should satisfy the

Helmholtz equations:

$$(\nabla^2+\mu_1^2)\underline{T}_1^* = -\varkappa\underline{X}^*, \quad (\nabla^2+\mu_2^2)\underline{T}_2^* = -\varkappa\underline{X}^*,$$
$$(\nabla^2+\mu_3^2)\underline{T}_3^* = -\varkappa\underline{X}^* \quad . \tag{2.3.66}$$

The solution of these equations is provided by the functions:

$$\underline{T}_j^*(\underline{x}) = \frac{\varkappa}{4\pi}\int_V \underline{X}^*(\underline{\xi})\frac{e^{ik_jR}}{R(\underline{\xi},\underline{x})}dV(\underline{\xi}), \quad j=1,2,3 . \tag{2.3.67}$$

Therefore

$$\underline{F}^*(\underline{x}) = \frac{\varkappa}{4\pi}\int_V \frac{\underline{X}^*(\underline{\xi})}{R(\underline{\xi},\underline{x})}\Big(\sum_{r=1}^{3}\frac{e^{i\mu_rR}}{D_rR}\Big)dV(\underline{\xi}) , \tag{2.3.68}$$

where

$$D_1 = \frac{1}{(\mu_1^2-\mu_2^2)(\mu_1^2-\mu_3^2)}, \quad D_2 = \frac{1}{(\mu_2^2-\mu_1^2)(\mu_2^2-\mu_3^2)}, \quad D_3 = \frac{1}{(\mu_3^2-\mu_1^2)(\mu_3^2-\mu_2^2)}.$$

Let us assume that a concentrated force $\underline{X}(\underline{x},t) = e^{-i\omega t}(X^*,0,0)$, where $\underline{X}^* = \delta(x_1)\,\delta(x_2)\,\delta(x_3)\,\delta_{1j}$ acts at the origin. Then, from Eq. (2.3.68), we shall obtain $\underline{F}^* = (F_1, 0,0)$, where

$$F_1^* = \frac{\varkappa}{4\pi R_0}\sum_{s=1}^{3}\frac{e^{i\mu_sR_0}}{D_s}, \qquad R_0 = (x_1^2+x_2^2+x_3^2)^{1/2}. \tag{2.3.69}$$

Upon substituting F_1^* of (2.3.69) in (2.3.63) and (2.3.64) and moving the concentrated force from the origin point $\underline{\xi}$ and directing it parallel to the x_l-axis, we obtain:

$$U_j^{*(l)}(\underline{x},\underline{\xi}) = \frac{\delta_{lj}}{4\pi R(\mu+\alpha)}\sum_{r=1}^{3}\frac{(\mu_3^2-\mu_r^2)(\sigma_4^2-2p-\mu_r^2)e^{i\mu_rR}}{D_r} +$$
$$-\frac{\lambda+\mu-\alpha}{4\pi(\lambda+2\mu)(\mu+\alpha)}\frac{\partial^2}{\partial x_l\partial x_j}\Big(\sum_{r=1}^{3}\frac{(\sigma_4^2-2p-\eta-\mu_r^2)e^{i\mu_rR}}{D_rR}\Big), \tag{2.3.70}$$

and

$$(2.3.71)\quad \Phi_j^{*(\ell)} = -\frac{2\alpha\epsilon_{jk\ell}}{4\pi(\mu+\alpha)(\gamma+\varepsilon)}\frac{\partial}{\partial x_k}\left(\sum_{r=1}^{3}\frac{\mu_3^2-\mu_r^2}{D_r R}e^{i\mu_r R}\right).$$

It can be shown that these equations are identical with (2.3.34) and (2.3.35). For this, use must be made of the relations:

$$\mu_1^2+\mu_2^2 = \sigma_2^2+\sigma_4^2+p(s-2)\quad,\qquad \mu_1^2\mu_2^2 = \sigma_2^2(\sigma_4^2-2p).$$

One of the fundamental theorems of the theory of elasticity is the theorem of reciprocity of works. For a body with micropolar elasticity, and if causes and effects vary with time in a harmonic manner, we have:

$$(2.3.72)\quad \int_V (X_i^* u_i'^* + Y_i^* \varphi_i'^*)\, dV = \int_V (X_i'^* u_i^* + Y_i'^* \varphi^*)\, dV .$$

In the form (2.3.72), the reciprocity theorem concerns, of course, an infinite body.

Let us consider loads of two types.

a. Let a concentrated force $X_j^* = \delta(\underline{x}-\underline{\xi})\delta_{jr}$ act at a point $\underline{\xi}$, thus producing a displacement $U_j^{*(r)}(\underline{x},\underline{\xi})$ field and a rotation field $\Phi_j^{*(r)}(\underline{x},\underline{\xi})$. Let now a concentrated force $X_j^{*\prime} = \delta(\underline{x}-\underline{\eta})\delta_{j\ell}$ act a point $\underline{\eta}$, parallel to the

x_ℓ -axis. This force will produce in the body a displacement $U_j^{*(\ell)}(\underline{x},\underline{\eta})$ and a rotation $\Phi_j^{*(\ell)}(\underline{x},\underline{\eta})$. From the reciprocity theorem (2.3.72), we have:

$$\int_V \delta(\underline{x}-\underline{\xi})\delta_{jr}U_j^{*(\ell)}(\underline{x},\underline{\eta})dV(\underline{x}) = \int_V \delta(\underline{x}-\underline{\eta})\delta_{j\ell}U_j^{*(r)}(\underline{x},\underline{\xi})dV(\underline{x}) .$$

Hence,

$$U_r^{*(\ell)}(\underline{\xi},\underline{\eta}) = U_\ell^{*(r)}(\underline{\eta},\underline{\xi}) . \tag{2.3.73}$$

b. Let a concentrated body couple $Y_j^* = \delta(\underline{x}-\underline{\xi})\delta_{jr}$ act at the point $\underline{\xi}$ and a body couple $Y_j'^* = \delta(\underline{x}-\underline{\eta})\delta_{j\ell}$ at the point $\underline{\eta}$. The body couple Y_j^* is connected with a field $V_j^{*(r)}$ and $W_j^{*(r)}$ and the body couple $Y_j'^*$ with a field $V_j^{*(\ell)}$ and $W_j^{*(\ell)}$. From Eq.(2.3.72) we obtain

$$V_r^{*(\ell)}(\underline{\eta},\underline{\xi}) = V_\ell^{*(r)}(\underline{\eta},\underline{\xi}) . \tag{2.3.74}$$

It can be seen, from (2.3.34) and (2.3.47), that the equations (2.3.73) and (2.3.74) are satisfied.

c. Let a concentrated force $X_j^* = \delta(\underline{x}-\underline{\xi})\delta_{jr}$ act at the point $\underline{\xi}$, thus producing a field $U_j^{*(r)}(\underline{x},\underline{\xi})$ and $\Phi_j^{*(r)}(\underline{x},\underline{\xi})$. Let now a concentrated body couple $Y_j^{*\prime} = \delta(\underline{x}-\underline{\eta})\delta_{j\ell}$ act at the point $\underline{\eta}$, in the direction of the x_ℓ-axis, thus producing a displacement field $V_j^{*(\ell)}(\underline{x},\underline{\eta})$ and a rotation field $W_j^{*(\ell)}(\underline{x},\underline{\eta})$.

From the reciprocity theorem (2.3.72), we

have:

$$\int_V \delta(\underline{x}-\underline{\xi})\delta_{jr}V_j^{*(\ell)}(\underline{x},\underline{\eta})dV(\underline{x}) = \int_V \delta(\underline{x}-\underline{\eta})\delta_{j\ell}\Phi_\ell^{*(r)}(\underline{x},\underline{\xi})dV(\underline{x}).$$

Hence,

$$V_r^{*(\ell)}(\underline{\xi},\underline{\eta}) = \Phi_\ell^{*}(\underline{\eta},\underline{\xi}). \tag{2.3.75}$$

By applying (2.3.35) and (2.3.47), we have:

$$\Phi_\ell^{*(r)}(\underline{\eta},\underline{\xi}) = \frac{p}{4\pi\rho c_2^2(k_1^2-k_2^2)}\epsilon_{r\ell k}\left|\frac{\partial}{\partial x_k}\left(\frac{e^{ik_1R}-e^{ik_2R}}{R(\underline{x},\underline{\xi})}\right)\right|_{\underline{x}=\underline{\eta}},$$

$$V_r^{*(\ell)}(\underline{\xi},\underline{\eta}) = \frac{s}{4\pi J c_4^2(k_1^2-k_2^2)}\epsilon_{\ell rk}\left|\frac{\partial}{\partial x_k}\left(\frac{e^{ik_1R}-e^{ik_2R}}{R(\underline{x},\underline{\eta})}\right)\right|_{\underline{x}=\underline{\xi}}.$$

It is evident, bearing in mind that $s = \frac{2\alpha}{\rho c_2^2}$, $p = \frac{2\alpha}{J c_4^2}$, that the relation (2.3.75) is satisfied.

2. 4 Singular Solutions of Higher Order.

Let us consider first the tri-dimensional problem. Let a concentrated force of intensity $-\frac{P}{\Delta\xi_1}e^{-i\omega t}$ act at the point $(\xi_1+\frac{\Delta\xi_1}{2}, \xi_2, \xi_3)$ parallel to the x_1-axis, and let a force of the same intensity act at the point $(\xi_1-\frac{\Delta\xi_1}{2}, \xi_2, \xi_3)$ in the direction of the positive x_1-axis. Then, the amplitude of the displacement u_j^* produced by these forces will be:

$$u_j^* = -\frac{P}{\Delta\xi_1}\left[U_j^*(x_1,x_2,x_3;\xi_1+\frac{\Delta\xi_1}{2},\xi_2,\xi_3) - U_j^*(x_1,x_2,x_3;\xi_1-\frac{\Delta\xi_1}{2},\xi_2,\xi_3)\right]. \tag{2.4.1}$$

By letting $\Delta\xi_1 \to 0$, we shall obtain the displacement $\hat{U}_j^{*(1)}$

for what is referred to as a double force without moment:

$$\hat{U}_j^{*(1)} = -P\frac{\partial}{\partial \xi_1} U_j^{*(1)}(\underline{x},\underline{\xi}). \tag{2.4.2}$$

Similarly, for the double force, we obtain the following rotation function

$$\hat{\Phi}_j^{*(1)} = -P\frac{\partial}{\partial \xi_1} \Phi_j^{*(1)}(\underline{x},\underline{\xi}). \tag{2.4.3}$$

Generally, if a double force without moment acts at the point $\underline{\xi}$ in the direction of the x_ℓ-axis, the corresponding singularities are given by the equations:

$$\hat{U}_j^{*(\ell)} = -P\frac{\partial}{\partial \xi_\ell} U_j^{*(\ell)}(\underline{x},\underline{\xi}), \tag{2.4.4}$$

$$\hat{\Phi}_j^{*(\ell)} = -P\frac{\partial}{\partial \xi_\ell} \Phi_j^{*(\ell)}(\underline{x},\underline{\xi}), \tag{2.4.5}$$

with the functions $U_j^{*(\ell)}$, $\Phi_j^{*(\ell)}$ as expressed by Eqs.(2.3.34) and (2.3.35).

Let now three double forces of intensity $Pe^{-i\omega t}$ act in the direction of the x_1, x_2 and x_3-axis. It is known that such a set of forces constitutes what is termed centre of compression or nucleus of dilatation. Let us denote by $\bar{U}_j^*$ the displacement components, and by $\bar{\Phi}^*$ the rotation components. Making use of the results obtained for double forces, we shall obtain, by superposition, the expression:

$$\bar{U}_j^* = -\left(\frac{\partial}{\partial \xi_1} U_j^{*(1)} + \frac{\partial}{\partial \xi_2} U_j^{*(2)} + \frac{\partial}{\partial \xi_3} U_j^{*(3)}\right) = \frac{1}{4\pi\rho c_1^2}\frac{\partial}{\partial \xi_j}\left(\frac{e^{i\sigma_1 R}}{R}\right) \tag{2.4.6}$$

(2.4.7) $$\bar{\Phi}_j^* = 0 .$$

It can easily be shown that a compression centre produces only longitudinal waves. Let a force $\frac{M}{\Delta\xi_1}e^{-i\omega t}$ act at the point $(\xi_1+\frac{\Delta\xi_1}{2},\xi_2,\xi_3)$ in the direction of the negative x_2-axis and let the same force at the point $(\xi_1-\frac{\Delta\xi_1}{2},\xi_2,\xi_3)$ in the opposite direction. Then,

$$u_j^* = \frac{M}{\Delta\xi_1}\left[U_j^{*(2)}(x_1,x_2,x_3;\xi_1-\frac{\Delta\xi_1}{2},\xi_2,\xi_3) - U_j^{*(2)}(x_1,x_2,x_3;\xi_1+\frac{\Delta\xi_1}{2},\xi_2,\xi_3)\right].$$

By making $\Delta\xi_1$ tend to zero, we find the displacement u_j^*, corresponding to the double force with moment

(2.4.8) $$u_j^* = -M\frac{\partial}{\partial\xi_1}U_j^{*(2)} .$$

Let now a force $\frac{M}{\Delta\xi_2}e^{-i\omega t}$ act at the point $(\xi_1,\xi_2+\frac{\Delta\xi_2}{2},\xi_3)$ in the direction of the positive x_1-axis and let a force of the same intensity act at the point $(\xi_1,\xi_2-\frac{\Delta\xi_2}{2},\xi_3)$ in the opposite direction of the x_1-axis As a result, we obtain:

(2.4.9) $$u_j^* = M\frac{\partial}{\partial\xi_2}U_j^{*(1)} .$$

The sum of these two double forces with moment will produce the displacements:

(2.4.10) $$u_j^* = -M\left(\frac{\partial U^{*(2)}}{\partial\xi_1} - \frac{\partial U^{*(1)}}{\partial\xi_2}\right).$$

Similarly, we obtain:

$$\varphi_j^* = -M\left(\frac{\partial \Phi^{*(2)}}{\partial \xi_1} - \frac{\partial \Phi^{*(1)}}{\partial \xi_2}\right). \qquad (2.4.11)$$

Making use of Eqs.(2.3.34) and (2.3.35), we obtain:

$$\underline{u}^* = \frac{M}{4\pi\varrho\omega^2}\left[\frac{\partial F}{\partial \xi_2}, -\frac{\partial F}{\partial \xi_1}, 0\right], \qquad (2.4.12)$$

where

$$F = A_1 k_1^2 \frac{e^{ik_1R}}{R} + A_2 k_2^2 \frac{e^{ik_2R}}{R},$$

and

$$\varphi_j^* = -\frac{Mp}{4\pi\varrho c_2^2(k_1^2-k_2^2)}\left(\epsilon_{2jk}\frac{\partial}{\partial \xi_1} - \epsilon_{1jk}\frac{\partial}{\partial \xi_2}\right)\frac{\partial}{\partial x_k}\left(\frac{e^{ik_1R}-e^{ik_2R}}{R}\right) \qquad (2.4.13)$$

or

$$\varphi_j^* = \frac{Mp}{4\pi\varrho c_2^2(k_1^2-k_2^2)}\left[\frac{\partial}{\partial \xi_j}\frac{\partial}{\partial x_3}\Gamma^*(\underline{x}) - \delta_{j3}\frac{\partial}{\partial \xi_k}\frac{\partial}{\partial x_k}\Gamma^*(\underline{x})\right], \qquad (2.4.13')$$

where

$$\Gamma^*(\underline{x}) = \frac{e^{ik_1R}-e^{ik_2R}}{R}.$$

Let now a concentrated couple of intensity $\frac{\mathfrak{M}}{\Delta\xi_1}e^{-i\omega t}$ act at the point $\left(\xi_1+\frac{\Delta\xi_1}{2}, \xi_2, \xi_3\right)$ in the direction of the negative x_1-axis and let a concentrated couple of the same intensity act at the point $\left(\xi_1-\frac{\Delta\xi_1}{2}, \xi_2, \xi_3\right)$ in the direction of the positive x_1-axis.

The amplitude u_j^* resulting from these two

body couples is:

$$(2.4.14)\quad u_j^* = -\frac{\mathfrak{M}}{\Delta\xi_1}\left[V_j^{*(1)}\left(x_1,x_2,x_3;\xi_1+\frac{\Delta\xi_1}{2},\xi_2,\xi_3\right) - V_j^{*(1)}\left(x_1,x_2,x_3;\xi_1-\frac{\Delta\xi_1}{2},\xi_2,\xi_3\right)\right].$$

If $\Delta\xi_1$ tends to zero, we obtain the displacement $\hat{V}_j^{*(1)}$ for the double couple:

$$(2.4.15)\qquad \hat{V}_j^{*(1)} = -\mathfrak{M}\frac{\partial}{\partial\xi_1}V_j^{*(1)}.$$

Similarly, for the rotation function, we have

$$(2.4.16)\qquad \hat{W}_j^{*(1)} = -\mathfrak{M}\frac{\partial}{\partial\xi_1}V_j^{*(1)}$$

the functions $V_j^{*(1)}$ and $W_j^{*(1)}$ being given by (2.3.47) and (2.3.48). If, now, three double couples of intensity $\mathfrak{M}$ act in the directions of the x_1, x_2 and x_3-axis, then, by superposition, we find that

$$(2.4.17)\qquad \bar{V}_j^* = -\mathfrak{M}\left(\frac{\partial}{\partial\xi_1}V_j^{*(1)} + \frac{\partial}{\partial\xi_2}V_j^{*(2)} + \frac{\partial}{\partial\xi_3}V_j^{*(3)}\right) = 0 \quad ,$$

$$(2.4.18)\qquad \bar{W}_j^* = \frac{\mathfrak{M}}{4\pi J c_3^2 k_3^2}\frac{\partial}{\partial\xi_j}\left(\frac{e^{ik_3R}}{R}\right), \qquad j = 1,2,3 \ .$$

The action of the three double couples can be treated as that of centre of microrotation. It is of interest to observe that there is no displacement field.

Let us now consider Eqs.(2.4.12) and (2.4.13'). In the classical theory of elasticity, Eq.(2.4.12) is treated as a vector of displacement produced by the action of a concentrated moment acting at the origin and directed along the negative x_3-axis. By confronting this equation with (2.3.47), which takes now the form

$$u_j^* = -\frac{Ms}{4\pi J c_4^2(k_1^2-k_2^2)}\epsilon_{3jk}\frac{\partial}{\partial x_k}\left(\frac{e^{ik_1R}-e^{ik_2R}}{R}\right), \quad (2.4.19)$$

it is seen that the results are not in agreement. This results from the fact that in the micropolar theory of elasticity a concentrated body couple is a fundamental load, similarly to concentrated forces. The above problem has been analysed in detail by P.P. Teodorescu+) in the static case.

Our considerations are also valid for the two-dimensional problem. Let us consider the case of a linear centre of compression. Let us make use of Eqs.(2.4.6 - 7) which takes a somewhat different form:

$$\bar{U}_j^* = -\left(\frac{\partial}{\partial \xi_1}U_j^{*(1)} + \frac{\partial}{\partial \xi_2}U_j^{*(2)}\right); \qquad \bar{\Phi}_j^* = 0, \quad (2.4.20)$$

where the displacement vector is taken from Eq. (2.3.54). As a result, we find:

$$\bar{U}_j^*(x_1, x_2; \xi_1, \xi_2) = \frac{1}{4\pi\rho c_1^2}H_0^{(1)}(\sigma r), \qquad \bar{\Phi}_j^* = 0, \quad (2.4.21)$$

where

$$r = [(x_1-\xi_1)^2 + (x_2-\xi_2)^2]^{1/2}.$$

+) Teodorescu,P.P.: Bull. Acad. Polon. Sci. Sér. Sci. Techn. 15, No.1, (1967), 57.

2.5 Generation of Waves in an Infinite Micropolar Space.

At the preceding sect.(2.3 and 2.4) we dealt with the fundamental solutions due to concentrated body forces and moments varying harmonically in time. Now we shall determine the displacements $\underline{u}(\underline{x}, t)$ and rotations $\underline{\varphi}(\underline{x}, t)$ produced by a general distribution of the body forces and moments when the variation in time is arbitrary +). As here the wave equations, listed below, will constitute our point of departure

(2.5.1) $$\left(\nabla^2 - \frac{1}{c_1^2}\partial_t^2\right)\Phi + \frac{1}{c_1^2}\vartheta = 0,$$

(2.5.2) $$\left(\nabla^2 - \tau_0^2 - \frac{1}{c_3^2}\partial_t^2\right)\Sigma + \frac{1}{c_3^2}\sigma = 0,$$

(2.5.3) $$\left[\left(\nabla^2 - \frac{1}{c_2^2}\partial_t^2\right)\left(\nabla^2 - \nu_0^2 - \frac{1}{c_4^2}\partial_t^2\right) + \eta_0^2\nabla^2\right]\underline{\Psi} = \frac{s}{c_4^2}\operatorname{rot}\underline{\eta} - \frac{1}{c_2^2}\left(\nabla^2 - \nu_0^2 - \frac{1}{c_4^2}\partial_t^2\right)\underline{\chi},$$

(2.5.4) $$\left[\left(\nabla^2 - \frac{1}{c_2^2}\partial_t^2\right)\left(\nabla^2 - \nu_0^2 - \frac{1}{c_4^2}\partial_t^2\right) + \eta_0^2\nabla^2\right]\underline{H} = \frac{p}{c_2^2}\operatorname{rot}\underline{\chi} - \frac{1}{c_4^2}\left(\nabla^2 - \frac{1}{c_2^2}\partial_t^2\right)\underline{\eta}.$$

We have introduced the following symbols

$$\tau_0^2 = \frac{4\alpha}{\beta + 2\gamma}, \qquad \nu_0^2 = \frac{4\alpha}{\gamma + \varepsilon}, \qquad \eta_0^2 = \frac{4\alpha}{(\gamma + \varepsilon)(\mu + \alpha)},$$

$$s = \frac{2\alpha}{\mu + \alpha}, \qquad p = \frac{2\alpha}{\gamma + \varepsilon}.$$

The displacements and rotations are connected with the elastic potentials Φ, Σ, Ψ, H in the following way

(2.5.5) $$\underline{u} = \operatorname{grad}\Phi + \operatorname{rot}\underline{\Psi}, \quad \operatorname{div}\underline{\Psi} = 0,$$
$$\underline{\varphi} = \operatorname{grad}\Sigma + \operatorname{rot}\underline{H}, \quad \operatorname{div}\underline{H} = 0.$$

+) Nowacki, W. and W.K. Nowacki: The generation of waves in an infinite micropolar elastic solid. Proc.Vibr.Probl.10, 2 (1969),169.

The quantities σ , ϑ , $\underline{\chi}$, $\underline{\eta}$ are combined with the vectors of body forces and moments

$$\underline{X} = \rho(\mathrm{grad}\,\vartheta + \mathrm{rot}\,\underline{\chi}), \qquad \mathrm{div}\,\underline{\chi} = 0,$$
$$\underline{Y} = \mathcal{J}(\mathrm{grad}\,\sigma + \mathrm{rot}\,\underline{\eta}), \qquad \mathrm{div}\,\underline{\eta} = 0. \tag{2.5.6}$$

To solve the system of wave equations (2.5.1) - (2.5.4) we make use of the four-dimensional Fourier transformation defined by the formulae

$$\tilde{\Phi}(\xi_1,\xi_2,\xi_3,\tau) = \frac{1}{4\pi^2}\int_{E_4}\Phi(x_1,x_2,x_3,t)\exp[i(x_k\xi_k+\tau t)]\,dV,$$
$$\Phi(x_1,x_2,x_3,t) = \frac{1}{4\pi^2}\int_{W_4}\tilde{\Phi}(\xi_1,\xi_2,\xi_3,\tau)\exp[-i(x_k\xi_k+\tau t)]\,dW, \tag{2.5.7}$$

where $dV = dx_1dx_2dx_3dt$, while E_4 denotes the whole x_1, x_2, x_3, t - space; $dW = d\xi_1\,d\xi_2\,d\xi_3\,d\tau$, while W_4 stands for the whole ξ_1 , ξ_2 , ξ_3 , τ - space.

Now, exploiting the results

$$\frac{1}{4\pi^2}\int_{E_4}\left(\frac{\partial\Phi}{\partial x_j}, \frac{\partial^2\Phi}{\partial t^2}\right)\exp[i(x_k\xi_k+\tau t)]\,dV = -(i\xi_j, \tau^2)\tilde{\Phi},$$

we obtain, in virtue of Eqs.(2.5.2) - (2.5.4) the following Fourier-transforms:

$$\tilde{\Phi} = \frac{1}{c_1^2}\frac{\tilde{\vartheta}}{\underline{\xi}^2 - \sigma_1^2}$$
$$\tilde{\Sigma} = \frac{1}{c_3^2}\frac{\tilde{\sigma}}{\underline{\xi}^2 + \tau_0^2 - \sigma_3^2} \tag{2.5.8a}$$

(2.5.8b)
$$\tilde{\Psi}_j = \frac{1}{\Delta}\left[\frac{1}{c_2^2}(\underline{\xi}^2 + \nu_0^2 - \sigma_4^2)\tilde{\chi}_j - \frac{is}{c_4^2}\xi_k \epsilon_{jkl}\tilde{\eta}_l\right],$$
$$\tilde{H}_j = \frac{1}{\Delta}\left[\frac{1}{c_4^2}(\underline{\xi}^2 - \sigma_2^2)\tilde{\eta}_j - \frac{i\rho}{c_2^2}\xi_k \epsilon_{jkl}\tilde{\chi}_l\right].$$

The following notations have been introduced in the above formulae

$$\sigma_j = \frac{\tau}{c_j}, \quad j = 1,2,3,4, \quad \Delta = (\underline{\xi}^2 - \lambda_1^2)(\underline{\xi}^2 - \lambda_2^2),$$
$$\lambda_{1,2}^2 = \frac{1}{2}\left[\sigma_2^2 + \sigma_4^2 + \eta_0^2 - \nu_0^2 \pm \sqrt{(\sigma_4^2 - \sigma_2^2 + \eta_0^2 - \nu_0^2)^2 + 4ps\sigma_2^2}\right],$$
$$\underline{\xi}^2 = \xi_1^2 + \xi_2^2 + \xi_3^2 .$$

We perform now the four-dimensional Fourier transformation on the expressions (2.5.5):

(2.5.9)
$$\tilde{u}_j = -i\xi_j\tilde{\Phi} - i\xi_k \epsilon_{jkl}\tilde{\Psi}_l,$$
$$\tilde{\varphi}_j = -i\xi_j\tilde{\Sigma} - i\xi_k \epsilon_{jkl}\tilde{H}_l .$$

Introducing into these relations $\tilde{\Phi}$, $\tilde{\Sigma}$, $\tilde{\Psi}_j$, $\tilde{H}_j$, taking into account the relation

$$\epsilon_{ljk}\epsilon_{lmn} = \delta_{jm}\delta_{kn} - \delta_{jn}\delta_{km}$$

and bearing in mind that $\operatorname{div}\underline{\chi} = 0$, $\operatorname{div}\underline{\eta} = 0$, we obtain the following formulae

(2.5.10a) $$\tilde{u}_j = -\frac{i\xi_j\vartheta}{c_1^2(\underline{\xi}^2 - \sigma_1^2)} + \frac{1}{\Delta}\left[\frac{s}{c_4^2}\underline{\xi}^2\tilde{\eta}_j - \frac{i}{c_2^2}(\underline{\xi}^2 + \nu_0^2 - \sigma_4^2)\epsilon_{jkl}\xi_k\tilde{\chi}_l\right],$$

$$\tilde{\varphi}_j = -\frac{i\xi_j\tilde{\sigma}}{c_3^2(\underline{\xi}^2+\tau_0^2-\sigma_3^2)} + \frac{1}{\Delta}\left[\frac{p}{c_2^2}\underline{\xi}^2\tilde{\chi}_j - \frac{1}{c_4^2}(\underline{\xi}^2-\sigma_2^2)\epsilon_{jkl}\xi_k\tilde{\eta}_l\right]. \quad (2.5.10b)$$

Let us apply a similar procedure to the formulae (2.5.6). We get

$$\begin{aligned}\tilde{X}_j &= -\rho\,(i\xi_j\tilde{\vartheta} + i\xi_k\epsilon_{jkl}\tilde{\chi}_l)\\ \tilde{Y}_j &= -J\,(i\xi_j\tilde{\sigma} + i\xi_k\epsilon_{jkl}\tilde{\eta}_l).\end{aligned} \quad (2.5.11)$$

By solving the above system of algebraic equations we arrive at

$$\tilde{\vartheta} = \frac{i\xi_k\tilde{X}_k}{\rho\underline{\xi}^2}, \qquad \tilde{\sigma} = \frac{i\xi_k\tilde{Y}_k}{J\underline{\xi}^2},$$

$$\tilde{\chi}_j = -\frac{i}{\rho\underline{\xi}^2}\epsilon_{jkl}\xi_k\tilde{X}_l, \qquad \tilde{\eta}_j = -\frac{i}{J\underline{\xi}^2}\epsilon_{jkl}\xi_k\tilde{Y}_l.$$

By introducing the values thus obtained into the relations (2.5.10) and performing the inverse Fourier transformation according to Eq. (2.5.7) we obtain the general solution of the system of equations

$$\Box_2\underline{u} + (\lambda+\mu-\alpha)\,\mathrm{grad\,div}\,\underline{u} + 2\alpha\,\mathrm{rot}\,\underline{\varphi} + \underline{X} = 0, \quad (2.5.12')$$

$$\Box_4\underline{\varphi} + (\beta+\gamma-\varepsilon)\,\mathrm{grad\,div}\,\underline{\varphi} + 2\alpha\,\mathrm{rot}\,\underline{u} + \underline{Y} = 0 \quad (2.5.12'')$$

in the form of a quadruple integral, namely

$$u_j(x_1,x_2,x_3,t)=\frac{1}{4\pi^2}\int_{W_4}\Big\{\frac{\xi_j\xi_k\tilde{X}_k}{\rho c_1^2\underline{\xi}^2(\underline{\xi}^2-\sigma_1^2)}-\frac{1}{\Delta}\Big[\frac{\underline{\xi}^2+\nu_0^2-\sigma_4^2}{c_2^2\rho\underline{\xi}^2}(\xi_j\xi_k\tilde{X}_k+$$
$$-\underline{\xi}^2\tilde{X}_j)+\frac{is}{Jc_4^2}\epsilon_{jkl}\xi_k\tilde{Y}_l\Big]\Big\}\exp[-i(x_k\xi_k+\tau t)]dW,$$

(2.5.13)

$$\varphi_j(x_1,x_2,x_3,t)=\frac{1}{4\pi^2}\int_{W_4}\Big\{\frac{\xi_j\xi_k\tilde{Y}_k}{Jc_3^2\underline{\xi}^2(\underline{\xi}^2+\tau_0^2-\sigma_3^2)}-\frac{1}{\Delta}\Big[\frac{\underline{\xi}^2-\sigma^2}{Jc_4^2\underline{\xi}^2}(\xi_j\xi_k\tilde{Y}_k+$$
$$-\underline{\xi}^2\tilde{Y}_j)+\frac{ip}{\rho c_2^2}\epsilon_{jkl}\xi_k\tilde{X}_l\Big]\Big\}\exp[-i(x_k\xi_k+\tau t)]dW.$$

Thus, the displacement and the rotations being known, we are able to determine the strain tensor γ_{ji} and the curvature-twist tensor $\varkappa_{ji}$:

$$\gamma_{ji}=u_{i,j}-\epsilon_{kji}\varphi_k\quad,\qquad \varkappa_{ji}=\varphi_{i,j}\ ,$$

and also the stress tensor σ_{ji} and couple tensor μ_{ji}

$$\sigma_{ji}=(\mu+\alpha)\gamma_{ji}+(\mu-\alpha)\gamma_{ij}+\lambda\gamma_{kk}\delta_{ji}\,,$$

(2.5.14)

$$\mu_{ji}=(\gamma+\varepsilon)\varkappa_{ji}+(\gamma-\varepsilon)\varkappa_{ij}+\beta\varkappa_{kk}\delta_{ji}\ .$$

Let us now consider the particular case when $\alpha\to 0$. Eqs. (2.5.12') and (2.5.12") become then independent from each other

$$\mu\nabla^2\underline{u}+(\lambda+\mu)\operatorname{grad}\operatorname{div}\underline{u}+\underline{X}=\rho\ddot{\underline{u}}$$

$$(\gamma+\varepsilon)\nabla^2\underline{\varphi}+(\gamma+\beta-\varepsilon)\operatorname{grad}\operatorname{div}\underline{\varphi}+\underline{Y}=J\ddot{\underline{\varphi}}\ .$$

Eqs. (2.5.14) are equations of classical elastokinetics; Eqs. $(2.5.14)_2$ refer to a hypothetical elastic medium where in only rotations are possible. In such a limit case we have in virtue of (2.5.13) the following formulae

$$u_j = \frac{1}{4\pi^2\mu}\int_{W_4} \frac{\tilde{X}_j\left(\delta^2\underline{\xi}^2 - \frac{\tau^2}{\hat{c}_2^2}\right) - (\delta^2 - 1)\xi_j\xi_k\tilde{X}}{\left(\underline{\xi}^2 - \frac{\tau^2}{\hat{c}_2^2}\right)\left(\underline{\xi}^2\delta^2 - \frac{\tau^2}{\hat{c}_2^2}\right)} \exp\left[-i(\xi_k x_k + \tau t)\right] dW,$$

$$\varphi_j = \frac{1}{4\pi^2(\gamma + \varepsilon)}\int_{W_4} \frac{\tilde{Y}_j\left(\rho_0^2\underline{\xi}^2 - \frac{\tau^2}{c_4^2}\right) - (\rho_0^2 - 1)\xi_j\xi_k\tilde{Y}}{\left(\underline{\xi}^2 - \frac{\tau^2}{c_4^2}\right)\left(\underline{\xi}^2\rho_0^2 - \frac{\tau^2}{c_4^2}\right)} \exp\left[-i(\xi_k x_k + \tau t)\right] dW. \tag{2.5.15}$$

Here

$$\delta^2 = \frac{\lambda + 2\mu}{\mu}\ , \quad \rho_0^2 = \frac{\beta + 2\gamma}{\gamma + \varepsilon}\ , \quad \hat{c}_2 = \left(\frac{\mu}{\rho}\right)^{1/2} .$$

In the static problem the body forces and the body couples do not depend on time. Denoting by $G_j(x_1, x_2, x_3)$ the components of body forces and by $M_j(x_1, x_2, x_3)$ the components of body couples we may express the transform $\tilde{X}_j$ as

$$\tilde{X}_j(\xi_1, \xi_2, \xi_3, \tau) = \frac{1}{4\pi^2}\int_{B_3} G_j(x_1, x_2, x_3)\exp(ix_k\xi_k)dA\int_{-\infty}^{\infty} e^{i\tau t}dt,$$

$$\int_{-\infty}^{\infty} e^{i\tau t}dt = 2\pi\delta(\tau) . \tag{2.5.16}$$

It results therefore that

$$\tilde{X}_j(\xi_1, \xi_2, \xi_3, \tau) = \sqrt{2\pi}\ \delta(\tau)\tilde{G}_j(\xi_1, \xi_2, \xi_3) \tag{2.5.17}$$

where

$$\tilde{G}_j(\xi_1, \xi_2, \xi_3) = \frac{1}{(2\pi)^{3/2}}\int_{B_3} G_j(x_1, x_2, x_3)\exp(ix_k\xi_k)dA.$$

Here $dA = dx_1dx_2dx_3$ while B_3 represents the

whole x_1, x_2, x_3-space.

Introducing into Eqs. (2.5.13) and performing the appropriate integration we obtain

$$u_j = \frac{1}{(2\pi)^{3/2}} \int_{D_3} \left\{ \frac{\xi_j \xi_k \tilde{G}_k}{\rho c_1^2 \underline{\xi}^2 \underline{\xi}^2} - \frac{1}{\Delta_0} \left[\frac{\underline{\xi}^2 + \nu_0^2}{c_2^2 \rho \underline{\xi}^2} (\xi_j \xi_k \tilde{G}_k + \right.\right.$$

$$(2.5.18) \qquad \left.\left. - \underline{\xi}^2 \tilde{G}_j) + \frac{i s}{J c_4^2} \epsilon_{jkl} \xi_k \tilde{M}_l \right] \right\} \exp(-i \xi_k x_k) \, dD ,$$

$$\varphi_j = \frac{1}{(2\pi)^{3/2}} \int_{D_3} \left\{ \frac{\xi_j \xi_k \tilde{M}_k}{J c_3^2 \underline{\xi}^2 (\underline{\xi}^2 + \tau_0^2)} - \frac{1}{\Delta_0} \left[\frac{1}{J c_4^2} (\xi_j \xi_k \tilde{M}_k + \right.\right.$$

$$(2.5.19) \qquad \left.\left. - \underline{\xi}^2 \tilde{M}_j) + \frac{i p}{\rho c_2^2} \epsilon_{jkl} \xi_k \tilde{G}_l \right] \right\} \exp(-i \xi_k x_k) \, dD ,$$

$$\underline{\xi}^2 = \xi_1^2 + \xi_2^2 + \xi_3^2 , \quad \Delta_0 = \underline{\xi}^2 (\underline{\xi}^2 + \nu_0^2 - \eta_0^2), \quad j, k, l = 1, 2, 3 .$$

Here $dD = d\xi_1 d\xi_2 d\xi_3$, while D_3 stands for the whole ξ_1, ξ_2, ξ_3-space.

Passing to the classical elastokinetics ($\alpha \to 0$) we obtain from (2.5.18) the following formula

$$(2.5.20) \qquad u_j = \frac{1}{(2\pi)^{3/2} \mu} \int_{D_3} \left(\frac{\tilde{G}_j}{\underline{\xi}^2} - \frac{(\delta^2 - 1) \xi_j \xi_k \tilde{G}_k}{\delta^2 \underline{\xi}^2 \underline{\xi}^2} \right) \exp(-i \xi_k x_k) \, dD .$$

By substituting $\alpha \to 0$ into Eq. (2.5.19) we get

$$(2.5.21) \qquad \varphi_j = \frac{1}{(2\pi)^{3/2} (\gamma + \varepsilon)} \int_{D_3} \left(\frac{\tilde{M}_j}{\underline{\xi}^2} - \frac{(\rho_0^2 - 1) \xi_j \xi_k \tilde{M}_k}{\rho_0^2 \underline{\xi}^2 \underline{\xi}^2} \right) \exp(-i \xi_k x_k) \, dD .$$

Let us consider a particular case. We shall assume that in the origin of the system of coordinates a concentrated force acts along the axis x_1 .

$$G_j = P_0\delta(x_1)\delta(x_2)\delta(x_3)\delta_{1j}\,, \qquad M_j = 0\,. \tag{2.5.22}$$

From the formulae (2.5.18) and (2.5.19) we obtain:

$$u_j^{(1)} = \frac{P_0}{8\pi^3}\int_{D_3}\left(\frac{\xi_j\xi_1}{\rho c_1^2\underline{\xi}^2\underline{\xi}^2} - \frac{(\underline{\xi}^2+\nu_0^2)(\xi_1\xi_j-\underline{\xi}^2\delta_{1j})}{\underline{\xi}^2\underline{\xi}^2(\underline{\xi}^2+\nu_0^2-\eta_0^2)\rho c_2^2}\right)\exp(-ix_k\xi_k)dD\,,$$

$$\varphi_j^{(1)} = -\frac{P_0 i p\,\epsilon_{jk1}}{8\pi^3\rho c_2^2}\int_{D_3}\frac{\xi_k\exp(-i\xi_k x_k)}{\xi^2(\xi^2+\nu_0^2-\eta_0^2)}\,dD\,. \tag{2.5.23}$$

After performing the integration and bearing in mind that

$$\int_{D_3}\frac{\exp(-ix_k\xi_k)}{\underline{\xi}^2-k_j^2}dD = 2\pi^2\frac{e^{ik_jR}}{R}\,,\quad \int_{D_3}\frac{\exp(-ix_k\xi_k)}{\underline{\xi}^2\underline{\xi}^2}\,dD = -\pi^2R, \tag{2.5.24}$$

we obtain the following expressions for displacements and rotations caused by the action of the concentrated force (2.5.22):

$$u_j^{(1)} = -\frac{\lambda+\mu}{8\pi\mu(\lambda+2\mu)}\partial_1\partial_j\left[R+\frac{(\gamma+\varepsilon)(\lambda+2\mu)}{2\mu(\lambda+\mu)}\left(\frac{1-e^{-R/l}}{R}\right)\right]+$$
$$-\frac{1}{4\pi\mu}\left(\frac{\alpha}{\mu+\alpha}\frac{e^{-R/l}}{R}-\frac{1}{R}\right)\delta_{1j}\,, \tag{2.5.25}$$

$$\varphi_j^{(1)} = \frac{1}{8\pi\mu}\,\epsilon_{1jk}\frac{\partial}{\partial x_k}\left(\frac{1-e^{-R/l}}{R}\right), \tag{2.5.26}$$

where

$$\frac{1}{l} = \left(\frac{4\alpha\mu}{(\mu+\alpha)(\gamma+\varepsilon)}\right)^{1/2}.$$

Those results are consistent with the solutions of N. Sandru+).

In the case in which in the origin of the system there acts a static concentrated couple with the vector oriented in the positive direction of the axis x_1 :

$$M_j = M_0\delta(x_1)\delta(x_2)\delta(x_3)\delta_{1j} \,, \qquad G_j = 0 \,. \tag{2.5.27}$$

Then for the displacements and rotations, we obtain from (2.5.18) and (2.5.19) the following expressions:

$$u_j^{(1)} = -\frac{M_0 i s \epsilon_{jk1}}{(2\pi)^3 J c_4^2}\int_{D_3}\frac{\exp(-ix_k\xi_k)}{\xi^2(\xi^2+\nu_0^2-\eta_0^2)}\,dD \,, \tag{2.5.28}$$

$$\varphi_j^{(1)} = \frac{M_0}{(2\pi)^3}\int_{D_3}\Big\{\frac{\xi_j\xi_1}{Jc_3^2\xi^2(\xi^2+\tau_0^2)} +$$

$$-\frac{1}{\Delta_0}\Big[\frac{1}{Jc_4^2}(\xi_j\xi_1-\xi^2\delta_{1j})\Big]\Big\}\exp(-ix_k\xi_k)\,dD \,, \tag{2.5.29}$$

from which we obtain upon integration:

$$u_j^{(1)} = -\frac{M_0}{8\pi\mu}\epsilon_{1jk}\frac{\partial}{\partial x_k}\left(\frac{e^{-R/l}-1}{R}\right), \tag{2.5.30}$$

$$\varphi_j^{(1)} = \frac{M_0}{16\pi\mu}\partial_1\partial_j\left[\frac{1-e^{-R/l}}{R}+\frac{\mu}{\alpha}\left(\frac{e^{-R/h}-e^{-R/l}}{R}\right)\right]+\frac{e^{-R/l}\delta_{1j}}{4\pi(\gamma+\varepsilon)} \,, \tag{2.5.31}$$

$$\frac{1}{h} = \left(\frac{4\alpha}{Jc_3^2}\right)^{1/2} .$$

+) Sandru, N. op. cit. p. 53.

When $\alpha \to 0$, then from (2.5.25), we obtain the solution for the classical theory of elasticity for the action of a static force concentrated in the origin of the system of coordinates in the direction of the x_1 axis:

$$u_i^{(1)} = \frac{\lambda + 3\mu}{8\pi\mu(\lambda+2\mu)}\left(\frac{1}{R}\delta_{1j} + \frac{\lambda+\mu}{\lambda+3\mu}\frac{x_1 x_j}{R^3}\right), \quad \varphi_i^{(1)} = 0. \quad (2.5.32)$$

2. 6 Propagation of Monochromatic Waves in an Infinite Micropolar Elastic Plate.

Let us now consider an elastic plate - we assume its thickness to be $2h$ - wherein a monochromatic wave propagates along the x_2-axis. We assume that the edges of the layer $x = \pm h$ are free of stresses. The following conditions should be satisfied on these edges

$$\sigma_{11} = 0, \ \sigma_{12} = 0, \ \mu_{13} = 0 \qquad \text{for} \qquad x_1 = \pm h. \quad (2.6.1)$$

Let us now consider a particular case of the equations

$$(\mu+\alpha)\Box_2 \underline{u} + (\lambda+\mu-\alpha)\,\mathrm{grad\,div}\,\underline{u} + 2\alpha\,\mathrm{rot}\,\underline{\varphi} = 0,$$
$$(2.6.2)$$
$$(\gamma+\varepsilon)\Box_4 \underline{\varphi} + (\beta+\gamma-\varepsilon)\,\mathrm{grad\,div}\,\underline{\varphi} + 2\alpha\,\mathrm{rot}\,\underline{u} = 0,$$

where the vectors $\underline{u}$ and $\underline{\varphi}$ are functions only of variable x_1, x_2 and time t. In this case we can derive from (2.6.2) two systems of equations independent of each other

$$(\mu+\alpha)\nabla_1^2 u_1 + (\mu-\alpha+\lambda)\partial_1 e + 2\alpha\partial_2\varphi_3 = \varrho\ddot{u}_1 ,$$

(2.6.3) $$(\mu+\alpha)\nabla_1^2 u_2 + (\mu+\lambda-\alpha)\partial_2 e - 2\alpha\partial_1\varphi_3 = \varrho\ddot{u}_2 ,$$

$$(\gamma+\varepsilon)\nabla_1^2\varphi_3 - 4\alpha\varphi_3 + 2\alpha(\partial_1 u_2 - \partial_2 u_1) = \mathcal{J}\ddot{\varphi}_3 ,$$

$$(\gamma+\varepsilon)\nabla_1^2\varphi_1 - 4\alpha\varphi_1 + (\gamma+\beta-\varepsilon)\partial_1\varkappa + 2\alpha\partial_2 u_3 = \mathcal{J}\ddot{\varphi}_1 ,$$

(2.6.4) $$(\gamma+\varepsilon)\nabla_1^2\varphi_2 - 4\alpha\varphi_2 + (\gamma+\beta-\varepsilon)\partial_2\varkappa - 2\alpha\partial_1 u_3 = \mathcal{J}\ddot{\varphi}_2 ,$$

$$(\mu+\alpha)\nabla_1^2 u_3 + 2\alpha(\partial_1\varphi_2 - \partial_2\varphi_1) = \varrho\ddot{u}_3 .$$

Here we have: $\nabla_1^2 = \partial_1^2 + \partial_2^2$, $e = \partial_1 u_1 + \partial_2 u_2$, $\varkappa = \partial_1\varphi_1 + \partial_2\varphi_2$.

The displacement and rotation field $\underline{u} = (u_1, u_2, 0)$, $\underline{\varphi} = (0,0,\varphi_3)$ described by Eqs. (2.6.3) induces the following stress $\underline{\sigma}$ and couple-stress $\underline{\mu}$ state

(2.6.5) $$\underline{\sigma} = \begin{vmatrix} \sigma_{11} & \sigma_{12} & 0 \\ \sigma_{21} & \sigma_{22} & 0 \\ 0 & 0 & \sigma_{33} \end{vmatrix} , \quad \underline{\mu} = \begin{vmatrix} 0 & 0 & \mu_{13} \\ 0 & 0 & \mu_{23} \\ \mu_{31} & \mu_{32} & 0 \end{vmatrix} ,$$

where

$$\sigma_{11} = 2\mu\partial_1 u_1 + \lambda e , \quad \sigma_{22} = 2\mu\partial_2 u_2 + \lambda e , \quad \sigma_{33} = \lambda e ,$$

(2.6.6) $$\sigma_{12} = \mu(\partial_1 u_2 + \partial_2 u_1) + \alpha(\partial_1 u_2 - \partial_2 u_1) - 2\alpha\varphi_3 ,$$

$$\sigma_{21} = \mu(\partial_1 u_2 + \partial_2 u_1) - \alpha(\partial_1 u_2 - \partial_2 u_1) + 2\alpha\varphi_3 ,$$

$$\mu_{13} = (\gamma + \varepsilon)\partial_1\varphi_3 \quad , \qquad \mu_{31} = (\gamma - \varepsilon)\partial_1\varphi_3 ,$$

$$\mu_{23} = (\gamma + \varepsilon)\partial_2\varphi_3 \quad , \qquad \mu_{32} = (\gamma - \varepsilon)\partial_2\varphi_3 .$$

As concerns the displacement and rotation field $\underline{u}=(0,0,u_3)$ and $\underline{\varphi}=(\varphi_1,\varphi_2,0)$ described by Eqs. (2.6.4), it induces the following stresses $\underline{\sigma}$ and couple stresses $\underline{\mu}$

$$\underline{\sigma} = \begin{vmatrix} 0 & 0 & \sigma_{13} \\ 0 & 0 & \sigma_{23} \\ \sigma_{31} & \sigma_{32} & 0 \end{vmatrix} , \quad \underline{\mu} = \begin{vmatrix} \mu_{11} & \mu_{12} & 0 \\ \mu_{21} & \mu_{22} & 0 \\ 0 & 0 & \mu_{33} \end{vmatrix} , \qquad (2.6.7)$$

where

$$\sigma_{13} = (\mu+\alpha)\partial_1 u_3 + 2\alpha\varphi_2 , \; \sigma_{31} = (\mu-\alpha)\partial_1 u_3 - 2\alpha\varphi_2 ,$$

$$\sigma_{23} = (\mu+\alpha)\partial_2 u_3 - 2\alpha\varphi_1 , \; \sigma_{32} = (\mu-\alpha)\partial_2 u_3 + 2\alpha\varphi_1 ,$$

$$\mu_{11} = 2\gamma\partial_1\varphi_1 + \beta\varkappa , \; \mu_{22} = 2\gamma\partial_2\varphi_2 + \beta\varkappa , \; \mu_{33} = \beta\varkappa , \qquad (2.6.8)$$

$$\mu_{12} = \gamma(\partial_1\varphi_2 + \partial_2\varphi_1) + \varepsilon(\partial_1\varphi_2 - \partial_2\varphi_1) ,$$

$$\mu_{21} = \gamma(\partial_1\varphi_2 + \partial_2\varphi_1) - \varepsilon(\partial_1\varphi_2 - \partial_2\varphi_1) .$$

We shall show that the system of Eqs. (2.6.3) leads to monochromatic waves known in the classical elastokinetics as Lamb's Wave.Eqs. (2.6.4) lead to the waves of Love's type.

a) Modified Lamb's Waves

Expressing the displacements by the potentials Φ, Ψ:

$$u_1 = \partial_1\Phi - \partial_2\Psi , \qquad u_2 = \partial_2\Phi + \partial_1\Psi , \qquad \varphi = \varphi_3 , \quad (2.6.9)$$

we can derive (putting $\underline{X} = \underline{Y} = 0$) from the system of Eqs. (2.6.3) the following equations

$$(2.6.10)\quad \begin{gathered}(\lambda + 2\mu)\nabla_1^2\Phi - \rho\ddot{\Phi} = 0, \quad (\mu+\alpha)\nabla_1^2\Psi - \rho\ddot{\Psi} - 2\alpha\varphi = 0,\\ [(\gamma+\varepsilon)\nabla_1^2 - 4\alpha - J\partial_t^2]\varphi + 2\alpha\nabla_1^2\Psi = 0.\end{gathered}$$

By eliminating from the last two equations first the quantity Ψ and then φ, we have

$$(2.6.11)\quad \{[(\mu+\alpha)\nabla_1^2 - \rho\partial_t^2][(\gamma+\varepsilon)\nabla_1^2 - 4\alpha - J\partial_t^2] + 4\alpha^2\nabla_1^2\}(\Psi,\varphi) = 0.$$

Eq.$(2.6.10)_1$ describes the longitudinal wave, while Eq.$(2.6.10)_2$ the modified transverse waves.

The solutions of Eqs.(2.6.10) and (2.6.11) will be sought for in the form

$$(2.6.12)\quad (\Phi, \Psi, \varphi) = (\Phi^*(x_1), \Psi^*(x_1), \varphi^*(x_1))e^{i(kx_2-\omega t)}.$$

These solutions are as follows

$$(2.6.13)\quad \begin{aligned}&\Phi^* = A\,\mathrm{sh}\,\delta x_1 + B\,\mathrm{ch}\,\delta x_1, \qquad \delta = (k^2 - \sigma_1^2)^{1/2}\\ &\Psi^* = C\,\mathrm{sh}\lambda_1 x_1 + D\,\mathrm{ch}\lambda_1 x_1 + E\,\mathrm{sh}\lambda_2 x_1 + F\,\mathrm{ch}\lambda_2 x_1,\\ &\varphi^* = C'\mathrm{sh}\lambda_1 x_1 + D'\mathrm{ch}\lambda_1 x_1 + E'\mathrm{sh}\lambda_2 x_1 + F'\mathrm{ch}\lambda_2 x_1.\end{aligned}$$

We introduced here the notation specified below

$$\sigma_1 = \frac{\omega}{c_1}, \; c_1 = \left(\frac{\lambda+2\mu}{\rho}\right)^{1/2}, \; \sigma_2 = \frac{\omega}{c_2}, \; c_2 = \left(\frac{\mu+\alpha}{\rho}\right)^{1/2},$$

and

$$\sigma_4 = \frac{\omega}{c_4}, \quad c_4 = \left(\frac{\gamma+\varepsilon}{J}\right)^{1/2}, \quad \nu_0^2 = \frac{4\alpha}{\gamma+\varepsilon}, \quad \eta_0^2 = \frac{4\alpha^2}{(\gamma+\varepsilon)(\mu+\alpha)}$$

$$\lambda_{1,2}^2 = k^2 + \frac{1}{2}\left(\nu_0^2 - \eta_0^2 - \sigma_2^2 - \sigma_4^2 \pm \sqrt{(\sigma_2^2 + \sigma_4^2 + \eta_0^2 - \nu_0^2)^2 - 4\sigma_2^2(\sigma_4^2 - \nu_0^2)}\right). \tag{2.6.13'}$$

Since the quantities λ_1^2 and λ_2^2 have to be positive (this follows from the postulate that the phase velocities be real), we have $\omega^2 > \frac{4\alpha}{J}$. The solutions $(2.6.13)_2$ and $(2.6.13)_3$ are connected through Eqs. $(2.6.10)_2$ and $(2.6.10)_3$, respectively.

Similarly, as in classical elastokinetics, the general problem of propagation of waves may be reduced to the solution of two simple problems, i.e. to the consideration of the symmetric and antisymmetric vibrations.

a. Symmetric vibrations are characterized by the symmetry of displacements u_2 and stresses σ_{11}, σ_{22} and μ_{13} with respect to the plane $x_1 = 0$. In this case we have to put in the expressions (2.6.13): $A = D = F = D' = F' = 0$. In view of the coupling of Eqs. $(2.6.10)_2$ and $(2.6.10)_3$, we have

$$C' = \varkappa_1 C_1 \quad , \quad E' = \varkappa_2 E \tag{2.6.14}$$

where

$$\varkappa_j = \frac{1}{s}(\sigma_2^2 + k^2 - \lambda_j^2), \quad j = 1,2, \qquad s = \frac{2\alpha}{\mu+\alpha}.$$

By expressing the boundary conditions (2.6.1) by the functions Φ^*, Ψ^* and φ^*, we obtain a system of three homogeneous equations. Making equal to zero the determinant of this system, we arrive at the following characteristic equation

$$\frac{\operatorname{tgh}(\delta h)}{\operatorname{tgh}(\lambda_1 h)} = \left(a_1\varkappa_2 - a_2\varkappa_1 \frac{\lambda_1}{\lambda_2}\frac{\operatorname{tgh}(\lambda_2 h)}{\operatorname{tgh}(\lambda_1 h)}\right)\frac{(2\mu+\lambda)\delta^2 - k^2\lambda}{(\varkappa_2-\varkappa_1)4\mu^2k^2\lambda_1\delta}, \tag{2.6.15}$$

where $a_j = \mu(k^2+\lambda_j^2) + \alpha(\lambda_j^2 - k^2) - 2\alpha\varkappa_j$, $j=1,2$. The quantity $c = \frac{k}{\omega}$ is the phase velocity sought for. From the transcendental Eq.(2.6.15) we obtain an infinite number of roots k. To each of these roots there corresponds a definite form of vibrations.

For $\alpha \to 0$ (what corresponds to the classical theory of elasticity) Eq.(2.6.15) reduces to the known transcendental equation for Lamb's waves +)

$$\frac{\operatorname{tgh}(kh\sqrt{1-c^2/c_1^2})}{\operatorname{tgh}(kh\sqrt{1-c^2/\hat{c}_2^2})} = \frac{\left(2-\frac{c^2}{\hat{c}_2^2}\right)^2}{4\sqrt{\left(1-\frac{c^2}{c_1^2}\right)\left(1-\frac{c^2}{\hat{c}_2^2}\right)}}, \quad \hat{c}_2 = \left(\frac{\mu}{\rho}\right)^{1/2}. \tag{2.6.16}$$

Let us now consider two particular cases. We assume first that the wavelength is small as compared with the thickness of the plate $2h$. Then the quantities δh, $\lambda_1 h$ and $\lambda_2 h$ are large such that it is plausible to assume the relation of hyperbolic tangents as equal to one. Then

$$\frac{\varkappa_2 a_1}{\varkappa_2-\varkappa_1} - \frac{a_2\lambda_1}{\lambda_2}\,\frac{\varkappa_1}{\varkappa_2-\varkappa_1} = \frac{4\mu^2k^2\lambda_1\delta}{(2\mu+\lambda)\delta^2 - k^2h}. \tag{2.6.17}$$

+) K.M. Ewing, W.S. Jardetzky, F. Press; Elastic waves in layered media. Mc Graw-Hill, New York, 1957.

The above equation coincides with the dispersional equation for the surface wave in a micropolar medium. For $\alpha \to 0$ we obtain from (2.6.17) the equation characteristic for Rayleigh waves+) :

$$\left(2-\frac{c^2}{\hat{c}^2}\right)^2 = 4\sqrt{\left(1-\frac{c^2}{c_1^2}\right)\left(1-\frac{c^2}{\hat{c}_2^2}\right)}. \quad (2.6.18)$$

For long waves, as compared with the thickness $2h$, the quantities $\delta h, \lambda_1 h, \lambda_2 h$ are small and the hyperbolic tangents in (2.6.15) may be replaced by their arguments. We have

$$4\mu^2 k^2 \delta(\varkappa_2 - \varkappa_1) = [(2\mu+\lambda)\delta^2 - k^2\lambda](a_1\varkappa_2 - a_2\varkappa_1). \quad (2.6.19)$$

In the particular case $\alpha \to 0$ there is

$$c = \frac{2\hat{c}_2}{c_1}(c_1^2 - \hat{c}_2^2)^{1/2} . \quad (2.6.20)$$

b. Antisymmetric vibrations. Let us now consider the particular case where the displacement u_2 and the stresses σ_{11}, σ_{22} and μ_{13} are antisymmetric with respect to the plane $x_1 = 0$. Then we have to put in the expressions (2.6.13): $B = C = E = C' = E' = 0$ and $D' = \varkappa_1 D$, $F' = \varkappa_2 F$.

Making use of the boundary conditions (2.6.1) we arrive at the transcendental equation

$$\left(\frac{a_1\varkappa_2\lambda_2}{\mathrm{tgh}(\lambda_1 h)} - \frac{a_2\varkappa_1\lambda_1}{\mathrm{tgh}(\lambda_2 h)}\right)\mathrm{tgh}(\delta h) = \frac{4\mu^2 k^2\delta^2\lambda_1\lambda_2(\varkappa_2-\varkappa_1)}{(2\mu+\lambda)\delta^2 - k^2\lambda}, \quad (2.6.21)$$

which permits to determine the successive values of the parame-

+) See footnote pag. 154.

ter k.

For $\alpha \to 0$ we obtain from Eq. (2.6.21) the transcendental equation of classical elastokinetics+)

$$\frac{\operatorname{tgh}\left(kh\sqrt{1-c^2/c_1^2}\right)}{\operatorname{tgh}\left(kh\sqrt{1-c^2/\hat{c}_2^2}\right)} = \frac{4\sqrt{\left(1-\frac{c^2}{c_1^2}\right)\left(1-\frac{c^2}{\hat{c}_2^2}\right)}}{\left(2-\frac{c^2}{\hat{c}_2^2}\right)^2} . \tag{2.6.22}$$

If the wavelength is very small compared with the thickness of the plate $2h$, Eq. (2.6.21) goes into (2.6.17). If, on the contrary, the length of the wave is large as compared with the thickness of the plate, then expanding the hyperbolic tangents into a series and retaining but two terms of the expanded form, we obtain the equation

$$\left(1-\frac{\delta^2 h^2}{3}\right)\left[\frac{a_1 \varkappa_2}{\lambda_1^2\left(1-\frac{\lambda_2^2 h^2}{3}\right)} - \frac{a_2 \varkappa_1}{\lambda_2^2\left(1-\frac{\lambda_1^2 h^2}{3}\right)}\right] = \frac{4\mu^2 k^2(\varkappa_2-\varkappa_1)}{(2\mu+\lambda)\delta^2 - k^2\lambda} . \tag{2.6.22'}$$

Therefore we are able to determine the phase velocity $c = \frac{\omega}{k}$ of the flexural wave. For $\alpha \to 0$ we obtain an expression known from the classical elastokinetics:

$$c^2 = \frac{4}{3}(kh)\hat{c}_2^2\left(1-\frac{\hat{c}_2^2}{c_1^2}\right) .$$

b) <u>The Modified Love's Waves</u>

Let us now consider an elastic plate $2h$ thick; the propagation of the monochromatic wave in such a medium is

---

+) See footnote pag. 154.

described by the system of Eqs. (2.6.4). We assume that the waves propagate with constant velocity along the x_2-axis. Then there is

$$(\varphi_1, \varphi_2, u_3) = (\varphi_1^*(x_1), \varphi_2^*(x_1), u_3^*(x_1))\, e^{i(kx_2 - \omega t)} . \quad (2.6.23)$$

By introducing into Eqs. (2.6.4) the potentials Σ and Ψ connected with the rotations φ_1, φ_2 by the relations

$$\varphi_1 = \partial_1\Sigma - \partial_2\Psi \quad , \qquad \varphi_2 = \partial_2\Sigma + \partial_1\Psi , \quad (2.6.24)$$

we separate these equations, obtaining the following system of equations

$$\begin{aligned} &[(\beta + 2\gamma)\nabla_1^2 - 4\alpha - J\partial_t^2]\,\Sigma = 0 , \\ &[(\gamma + \varepsilon)\nabla_1^2 - 4\alpha - J\partial_t^2]\,\Psi - 2\alpha u_3 = 0 , \\ &[(\mu + \alpha)\nabla_1^2 - \rho\partial_t^2]\,u_3 + 2\alpha\nabla_1^2\Psi = 0 . \end{aligned} \quad (2.6.25)$$

By eliminating from the two last equations first the quantity Ψ and then u_3 we get an equation identical as to its structure with Eq. (2.6.11)

$$\{[(\gamma + \varepsilon)\nabla_1^2 - 4\alpha - J\partial_t^2][(\mu + \alpha)\nabla_1^2 - \rho\partial_t^2] + 4\alpha^2\nabla_1^2\}(\Psi, u_3) = 0 . \quad (2.6.26)$$

Now, requiring the boundary of the plate to be free of stresses, we have the following boundary conditions

$$\mu_{11} = 0, \; \mu_{12} = 0, \; \sigma_{13} = 0 \quad \text{for} \quad x_1 = \pm h . \quad (2.6.27)$$

The first Eq. (2.6.25) represents the micro-rotational wave while the second and the third Eq. (2.6.25) describe the transverse modified waves.

The solutions of Eqs. (2.6.25) and (2.6.26) will be sought for in the form

$$\Sigma^* = A\,\mathrm{sh}\,\sigma x_1 + B\,\mathrm{ch}\,\sigma x_1\,,$$

$$\Psi^* = C\,\mathrm{sh}\lambda_1 x_1 + D\,\mathrm{ch}\lambda_1 x_1 + E\,\mathrm{sh}\lambda_2 x_1 + F\,\mathrm{ch}\lambda_2 x_1\,, \tag{2.6.28}$$

$$u_3^* = C'\mathrm{sh}\lambda_1 x_1 + D'\mathrm{ch}\lambda_1 x_1 + E'\mathrm{sh}\lambda_2 x_1 + F'\mathrm{ch}\lambda_2 x_1\,.$$

The following notations have been introduced into the above formulae

$$\sigma = (k^2 + \tau_0^2 - \sigma_3^2)^{1/2},\ \sigma_3 = \frac{\omega}{c_3}\,,\ c_3 = \left(\frac{\beta+2\gamma}{J}\right)^{1/2},\ \tau_0^2 = \frac{4\alpha}{\beta+2\gamma}\,.$$

The quantities λ_1, λ_2 are given by the formulae (2.6.13').

a. Symmetric vibrations. We require the rotation φ_2 and the stresses μ_{11}, μ_{22}, σ_{13} to be symmetric with respect to the plane $x_1 = 0$. This postulate will be satisfied if we assume $A = D = F = D' = F' = 0$ and $C' = \varrho_1 C$, $E' = \varrho_2 E$, where the quantities ϱ_1 and ϱ_2 may be determined from the third Eq. (2.6.25). Thus, we have

$$\varrho_j = \frac{s(k^2 - \lambda_j^2)}{\lambda_j^2 - k^2 + \sigma_2^2}\,,\quad s = \frac{2\alpha}{\mu+\alpha}\,,\qquad j = 1, 2\,. \tag{2.6.29}$$

By taking into account boundary conditions expressed by (2.6.27) we obtain the system of three homogeneous equations. By making equal to zero the determinant of this system, we get a transcendental equation as below

$$\frac{\mathrm{tgh}(\sigma h)}{\mathrm{tgh}(\lambda_1 h)} = \frac{4\gamma\alpha k^2\left(e_2\lambda_1\frac{\mathrm{tgh}(\lambda_2 h)}{\mathrm{tgh}(\lambda_1 h)} - e_1\lambda_2\right) + [(2\gamma+\beta)\sigma^2 - \beta k^2]\left(e_1 d_2 - e_2 d_1\frac{\mathrm{tgh}(\lambda_2 h)}{\mathrm{tgh}(\lambda_1 h)}\right)}{4k^2\gamma^2\sigma(\lambda_1 d_2 - \lambda_2 d_2)}, \tag{2.6.30}$$

where

$$e_j = \gamma(\lambda_j^2 + k^2) + \varepsilon(\lambda_j^2 - k^2),\quad d_j = (\mu+\alpha)\lambda_j\rho_j + 2\alpha\lambda_j,\quad j = 1,2 .$$

The successive values of the parameter $k = \frac{\omega}{c}$ and the corresponding phase velocities c and forms of waves may be determined from Eq.(2.6.30).

For small lengths waves, as compared with the thickness of the plate, we obtain the equation

$$4k^2\gamma^2\sigma(\lambda_1 d_1 - \lambda_2 d_2) = 4\gamma\alpha k^2(e_2 d_1 - e_1 d_2) + [(2\gamma+\beta)\sigma^2 - \beta^2 k^2](e_1 d_2 - e_2 d_1)\ , \tag{2.6.31}$$

wherefrom we may determine the phase velocity of the surface wave in an elastic half-space. Thus, it appears that in a micropolar elastic medium we have not only the waves of Rayleigh type but also the waves of Love type $u_3(x_1, x_2, t) = u_3^*(x_1)\, e^{i(kx_2 - \omega t)}$ accompanied by the modified twist waves φ_1, φ_2. In the classical medium the appearance of Love waves was possible only in a layered half-space provided certain definite inequalities concerning the material constants were satisfied+).

+) W.M. Ewing, W.S. Jardetzky, F. Press, loc. cit. p.154.

Let us return once more to Eq.(2.6.30), assuming $\alpha \to 0$. We obtain

$$(2.6.32)\quad \frac{\operatorname{tgh}(kh\sqrt{1-c^2/c_3^2})}{\operatorname{tgh}(kh\sqrt{1-c^2/c_4^2})} = \frac{(2-c^2/c_0^2)^2}{4\sqrt{(1-c^2/c_3^2)(1-c^2/c_4^2)}}\,,\quad c_0 = \left(\frac{\gamma}{J}\right)^{1/2}.$$

b. Antisymmetric vibrations. Let us consider the case, where the rotation φ_2 and the stresses μ_{11}, μ_{22}, σ_{13} are antisymmetric with respect to the plane $x_1 = 0$. By assuming in (2.6.28) $B = C = E = C' = E' = 0$ and $D' = \varrho_1 D$, $F' = \varrho F$, we obtain - taking into account the boundary conditions (2.6.27) - the following transcendental equation

$$\frac{\operatorname{tgh}(\sigma h)}{\operatorname{tgh}(\lambda_1 h)} =$$

$$(2.6.33) = -\frac{4k^2\gamma^2\sigma(\lambda_1 d_2 - d_1\lambda_2)}{[(2\gamma+\beta)\sigma^2-\beta k^2]\left[e_1 d_2 - d_1 e_2\frac{\operatorname{tgh}(\lambda_1 h)}{\operatorname{tgh}(\lambda_2 h)}\right] + 4\gamma\alpha^2 h\left[\lambda_1 e_2\frac{\operatorname{tgh}(\lambda_1 h)}{\operatorname{tgh}(\lambda_2 h)} - \lambda_2 e_1\right]}.$$

For very small lengths of waves, as compared with the thickness of the plate, we get from Eq.(2.6.33) again Eq.(2.6.31). In the particular case $\alpha \to 0$ we obtain the equation

$$(2.6.34)\quad \frac{\operatorname{tgh}(kh\sqrt{1-c^2/c_3^2})}{\operatorname{tgh}(kh\sqrt{1-c^2/c_4^2})} = \frac{4(1-c^2/c_3^2)^{1/2}(1-c^2/c_4^2)^{1/2}}{(2-c^2/c_0^2)^2}.$$

If in all transcendental equations referring to the modified Lamb's and Love's problem we put $k = 0$, these equations will refer to the free vibrations of the elastic layer which depend solely on x_1 and t. It means that they are monochromatic one dimensional vibrations.

Chapter 3.

Statical Problems of Micropolar Elasticity.

3. 1 Fundamental Relations and Equations

Let us apply to a body the loadings, increasing them slowly from zero to their final values. The heat content changes together with the deformation. Consequently, the field of temperature in the body also changes. We assume that in the course of deformation there exists a permanent heat exchange with the environment of the body, and, therefore, the process can be regarded to be isothermal.

In the steady state of deformation we assume that $T = T_0$ ($\vartheta = 0$) and that all field quantities, the displacement $\underline{u}$, and rotation $\underline{\varphi}$, are independent of time.

In the sequel we shall make use of the results given at section 1.3, assuming that the increment of temperature is equal to zero. Thus, the expression for the free energy takes the following form (formula 1.4.1):

$$F = \frac{\mu+\alpha}{2}\gamma_{ji}\gamma_{ji} + \frac{\mu-\alpha}{2}\gamma_{ji}\gamma_{ij} + \frac{\lambda}{2}\gamma_{kk}\gamma_{nn} + \frac{\gamma+\varepsilon}{2}\varkappa_{ji}\varkappa_{ji} + \frac{\gamma-\varepsilon}{2}\varkappa_{ji}\varkappa_{ij} + \frac{\beta}{2}\varkappa_{kk}\varkappa_{nn} . \tag{3.1.1}$$

Hence, taking into account the relations

$$\sigma_{ji} = \frac{\partial F}{\partial \gamma_{ji}} , \quad \mu_{ji} = \frac{\partial F}{\partial \varkappa_{ji}} \tag{3.1.2}$$

we obtain the following constitutive equations

$$\sigma_{ji} = (\mu + \alpha)\gamma_{ji} + (\mu - \alpha)\gamma_{ij} + \lambda\gamma_{kk}\delta_{ji}, \qquad (3.1.3)$$

$$\mu_{ji} = (\gamma + \varepsilon)\varkappa_{ji} + (\gamma - \varepsilon)\varkappa_{ij} + \beta\varkappa_{kk}\delta_{ji}. \qquad (3.1.4)$$

Here the material constants μ, λ, α, β, γ, ε refer to an isothermal process.

Let us consider the simple states of stresses. If a long cylinder, the axis of which is parallel to the x_1 axis, is uniformly extended, then the ratio $\frac{\sigma_{11}}{\gamma_{11}}$ defines the Young modulus E of its material.

By substituting $\sigma_{11} = E\gamma_{11}$, $\sigma_{22} = \sigma_{33} = 0$ in Eqs. (3.1.3) we obtain

$$\sigma_{11} = E\gamma_{11} = (\lambda + 2\mu)\gamma_{11} + \lambda(\gamma_{22} + \gamma_{33})$$

$$0 = (\lambda + 2\mu)\gamma_{22} + \lambda(\gamma_{11} + \gamma_{33})$$

$$0 = (\lambda + 2\mu)\gamma_{33} + \lambda(\gamma_{11} + \gamma_{22}).$$

Hence, we get

$$E = \frac{\mu(3\lambda + 2\mu)}{\lambda + 2\mu}.$$

We have expressed the Young modulus by the constants μ, λ. Transversal contraction $\nu = -\frac{\gamma_{22}}{\gamma_{11}} = -\frac{\gamma_{33}}{\gamma_{11}}$ can be expressed by the constants μ, λ in the following way

$$\nu = \frac{\lambda}{2(\lambda + \mu)}.$$

If the cylinder is subject to the action of a uniform torsion then the ratio $\frac{\mu_{11}}{\varkappa_{11}}$ defines the modulus $\overset{\circ}{E}$ of the material. By substituting $\mu_{11} = \overset{\circ}{E}\varkappa_{11}$, $\mu_{22} = \mu_{33} = 0$ in Eqs.(3.1.4) we obtain

$$\mu_{11} = \overset{\circ}{E}\varkappa_{11} = (\beta + 2\gamma)\varkappa_{11} + \beta(\varkappa_{22} + \varkappa_{33}),$$

$$0 = (\beta + 2\gamma)\varkappa_{22} + \beta(\varkappa_{11} + \varkappa_{33}),$$

$$0 = (\beta + 2\gamma)\varkappa_{33} + \beta(\varkappa_{11} + \varkappa_{22}).$$

Hence, we have

$$\overset{\circ}{E} = \frac{\gamma(3\beta + 2\gamma)}{\beta + 2\gamma}, \quad \overset{\circ}{\nu} = -\frac{\varkappa_{22}}{\varkappa_{11}} = -\frac{\varkappa_{22}}{\varkappa_{11}} = \frac{\beta}{2(\gamma + \beta)}.$$

Next, assume that a hydrostatic pressure acts, then $\sigma_{ji} = -p\delta_{ji}$, $p > 0$.

We obtain from the equation $\sigma_{kk} = (3\lambda + 2\mu)\gamma_{kk}$ that

$$\gamma_{kk} = -\frac{p}{K}, \quad K = \lambda + \frac{2}{3}\mu.$$

The volume of the body decreases. We recognize in $K > 0$ the modulus of compression (or the bulk modulus) known from the classical theory of elasticity.

Now, assume that the body is subject to a uniform, three-dimensional torsion, then $\mu_{ji} = -m\delta_{ji}$, $m > 0$. We find from the formula $\mu_{kk} = (3\beta + 2\gamma)\varkappa_{kk}$

$$\varkappa_{kk} = -\frac{m}{L}, \quad L = \beta + \frac{2}{3}\gamma.$$

The quantity L can be regarded as the measure of torsion of the body.

If we substitute Eqs.(3.1.3) and (3.1.4) in the equilibrium equations

$$\sigma_{ji,j} + X_i = 0 \quad , \qquad \epsilon_{ijk}\sigma_{jk} + \mu_{ji,j} + Y_i = 0, \qquad (3.1.5)$$

and make use of the definitions

$$\gamma_{ji} = u_{i,j} - \epsilon_{kji}\varphi_k \quad , \qquad \varkappa_{ji} = \varphi_{i,j} \qquad (3.1.6)$$

then, in the result, we obtain the system of six differential equations

$$(\mu+\alpha)\nabla^2\underline{u} + (\lambda+\mu-\alpha)\,\mathrm{grad\,div}\,\underline{u} + 2\alpha\,\mathrm{rot}\,\underline{\varphi} + \underline{X} = 0, \qquad (3.1.7)$$

$$(\gamma+\varepsilon)\nabla^2\underline{\varphi} - 4\alpha\underline{\varphi} + (\beta+\gamma-\varepsilon)\,\mathrm{grad\,div}\,\underline{\varphi} + 2\alpha\,\mathrm{rot}\,\underline{u} + \underline{Y} = 0. \qquad (3.1.8)$$

The equations have to be considered with the corresponding boundary conditions. Assume that on a part of the surface A, denoted by A_u, the displacements $\underline{u}$ and rotations $\underline{\varphi}$ be given. On the other part of the surface $A_\sigma = A - A_u$ the loadings $\underline{p}$ and the moments $\underline{m}$ are prescribed. Thus the boundary conditions take the following form

$$u_i = \hat{u}_i(\underline{x}) \,, \qquad \varphi_i = \hat{\varphi}_i(\underline{x}) \,, \qquad \underline{x} \in A_u, \qquad (3.1.9)$$

$$p_i = \sigma_{ji}n_j = \hat{p}_i(\underline{x}), \; m_i = \mu_{ji}n_j = \hat{m}_i(\underline{x}), \qquad \underline{x} \in A_\sigma. \qquad (3.1.10)$$

Here, a component of the unit vector, normal to the surface A, is denoted by $\underline{n}$, and the sense of the vector is outward of the surface A . The functions $\hat{u}_i$, $\hat{\varphi}_i$, $\hat{m}_i$, p_i are the functions given on the boundary.

3. 2 Theorem of Minimum of the Potential and the Complementary Energy.

Let the body be in the state of static equilibrium under the action of external forces. Let the components of the displacement vector $\underline{u}$ and of the rotation vector $\underline{\varphi}$ be given on the surface A_u , and tensions $\underline{p}$ and moments $\underline{m}$ - on A_σ .

Let us assume that there exists a system of displacements u_i and rotations φ_i satisfying the equalibrium equations. We shall consider the displacements $u_i + \delta u_i$ and rotations $\varphi_i + \delta\varphi_i$ consistent with the constraints imposed on the body. Virtual displacements δu_i and rotations $\delta\varphi_i$ ought to be the functions of the class $C^{(2)}$, taking zero values on A_u and arbitrary values on A_σ .

The virtual work principle takes, now, the form

$$(3.2.1)\quad \int_V (X_i\delta u_i + Y_i\delta\varphi_i)dV + \int_A (p_i\delta u_i + m_i\delta\varphi_i)dA = \int_V (\sigma_{ji}\delta\gamma_{ji} + \mu_{ji}\delta\varkappa_{ji})dV .$$

This equation may be transformed - taking into account the constitutive relations - as follows:

$$(3.2.2)\quad \int_V (X_i\delta u_i + Y_i\delta\varphi_i)dV + \int_A (p_i\delta u_i + m_i\delta\varphi_i)dA = \delta W_\varepsilon .$$

Here

$$d\mathcal{W}_\varepsilon = \int_V [2\mu\gamma_{(ij)}\delta\gamma_{(ij)} + 2\alpha\gamma_{\langle ij\rangle}\delta\gamma_{\langle ij\rangle} + 2\gamma\varkappa_{(ij)}\delta\varkappa_{(ij)} +$$
$$+ 2\varepsilon\varkappa_{\langle ij\rangle}\delta\varkappa_{\langle ij\rangle} + \lambda\gamma_{kk}\delta\gamma_{nn} + \beta\varkappa_{kk}\delta\varkappa_{nn}]\, dV.$$

Since the body forces and the body couples as well as the tensions and moments of surface do not vary, we may write Eq.(3.2.2) in the following form

$$\delta\Gamma = 0 \tag{3.2.3}$$

where

$$\Gamma = \mathcal{W}_\varepsilon - \int_V (X_i u_i + Y_i\varphi_i)\, dV - \int_{A_\sigma} (p_i u_i + m_i\varphi_i)\, dA. \tag{3.2.4}$$

The quantity Γ, called the potential energy, is extremum. By proceeding in an analogous way as for symmetric elasticity, we arrive at the conclusion that Γ is minimum. The theorem on potential energy states that from among all the displacements u_i and rotations φ_i which statisfy the given boundary conditions only those fulfilling at the same time the equilibrium equations lead to the minimum of potential energy.

Let us solve the constitutive relations with respect to γ_{ij} and $\varkappa_{ij}$. We have

$$\begin{aligned} \gamma_{ij} &= 2\mu'\sigma_{(ij)} + 2\alpha'\sigma_{\langle ij\rangle} + \lambda'\delta_{ij}\sigma_{kk}, \\ \varkappa_{ij} &= 2\gamma'\mu_{(ij)} + 2\varepsilon'\mu_{\langle ij\rangle} + \beta'\delta_{ij}\mu_{kk}. \end{aligned} \tag{3.2.5}$$

It is easy to prove that

$$\gamma_{ji} = \frac{\partial F}{\partial \sigma_{ji}} \qquad \varkappa_{ji} = \frac{\partial F}{\partial \mu_{ji}} \tag{3.2.6}$$

if F is expressed as the function of stresses σ_{ij} and couple-stresses μ_{ij}.

Here is

$$F = \mu' \sigma_{(ij)} \sigma_{(ij)} + \alpha' \sigma_{\langle ij \rangle} \sigma_{\langle ij \rangle} + \gamma' \mu_{(ij)} \mu_{(ij)} + \varepsilon' \mu_{\langle ij \rangle} \mu_{\langle ij \rangle} + \\ + \frac{\lambda'}{2} \sigma_{kk} \sigma_{nn} + \frac{\beta'}{2} \mu_{kk} \mu_{nn} \tag{3.2.7}$$

we shall consider the integral

$$I = \int_V \left(\gamma_{ji} \delta \sigma_{ji} + \varkappa_{ji} \delta \mu_{ji} \right) dV . \tag{3.2.8}$$

In this expression $\delta \sigma_{ji}$, $\delta \mu_{ji}$ denote the virtual increments of stresses and couple-stresses. These increments are regarded as functions of class $C^{(1)}$, as very small and arbitrary quantities. By taking into consideration (3.2.6), we have

$$\int_V (\gamma_{ji} \delta \sigma_{ji} + \varkappa_{ji} \delta \mu_{ji}) dV = \delta W_\sigma , \qquad W_\sigma = \int_V F \, dV \tag{3.2.9}$$

where

$$\delta W_\sigma = \int_V \left(\frac{\partial F}{\partial \sigma_{ji}} \delta \sigma_{ji} + \frac{\partial F}{\partial \mu_{ji}} \delta \mu_{ji} \right) dV . \tag{3.2.10}$$

By transforming the left-hand side of Eq.(3.2.9), taking into account the relation $\gamma_{ji} = u_{i,j} - \epsilon_{kji} \varphi_k$, $\varkappa_{ji} = \varphi_{i,j}$ and introducing notations $\delta p_i = \delta \sigma_{ji} n_j$, $\delta m_i = \delta \mu_{ji} n_j$ we obtain

$$\int_A (u_i\delta p_i + \varphi_i\delta m_i)dA - \int_V [u_i\delta\sigma_{ji,j} + \varphi_i(\epsilon_{ijk}\delta\sigma_{jk} + \delta\mu_{ji,j})]dV = \delta W_\sigma . \tag{3.2.11}$$

We require the stresses $\sigma_{ji} + \delta\sigma_{ji}$ and couple-stresses $\mu_{ji} + \delta\mu_{ji}$ to be statically possible. It means that the equilibrium conditions

$$\sigma_{ji,j} + \delta\sigma_{ji,j} + X_i + \delta X_i = 0 , \tag{3.2.12}$$

$$\epsilon_{ijk}(\sigma_{jk} + \delta\sigma_{jk}) + \mu_{ji,j} + \delta\mu_{ji,j} + Y_i + \delta Y_i = 0 , \tag{3.2.13}$$

have to be satisfied inside the volume V and the boundary conditions

$$p_i + \delta p_i = (\sigma_{ji} + \delta\sigma_{ji})n_j , \quad m_i + \delta m_i = (\mu_{ji} + \delta\mu_{ji})n_j \tag{3.2.14}$$

on the surface A_σ .

The quantities $\delta\sigma_{ji}$ and $\delta\mu_{ji}$ on A_u may be arbitrary. In view of the equilibrium equations and boundary conditions we have

$$\delta\sigma_{ji,j} + \delta X_i = 0 , \quad \epsilon_{ijk}\delta\sigma_{jk} + \delta\mu_{ji,j} + \delta Y_i = 0 , \qquad \underline{x} \in V,$$

and

$$\delta p_i = \delta\sigma_{ji}n_j , \qquad \delta m_i = \delta\mu_{ji}n_j , \qquad \underline{x} \in A_\sigma .$$

As we want to compare all the fields of stresses and couple-stresses satisfying the equilibrium equations, but not necessarily the compatibility equation, it should be assumed that $\delta X_i = 0$,

$\delta Y_i = 0$ inside the volume V, and $\delta p_i = 0$, $\delta Y_i = 0$ on the surface A_σ, leaving the increments δp_i, δm_i on the A_σ surface arbitrary. Under These restrictions Eq. (3.2.11) takes the form

$$\int_{A_u} (u_i \delta p_i + \varphi_i \delta m_i)\, dA = \delta \mathcal{W}_\sigma . \tag{3.2.15}$$

Because displacements u_i rotations φ_i do not vary, we have

$$\delta \Gamma^* = 0 \tag{3.2.16}$$

$$\Gamma^* = \mathcal{W}_\sigma - \int_{A_u} (p_i u_i + m_i \varphi_i)\, dA , \quad \mathcal{W}_\sigma = \int_V F\, dV . \tag{3.2.17}$$

The expression Γ^* is said to represent the complementary work. Similarly as in the theory of symmetric elasticity, it can be, here said, that Γ^* becomes minimum. Eq.(3.2.16) is the theorem on minimum of the complementary work extended to the problem of the theory of asymmetric elasticity. This theorem says that from among all the tensor fields σ_{ji}, μ_{ji}, satisfying the equilibrium equations and the boundary conditions given by the tensions p_i and moments m_i, only those actually occur which reduce the functional to minimum.

3. 3 The Extended Reissner's Theorem.

Thus, the Reissner's variational theorem +) formulated in most general terms can be easily extended so as to

+) E. Reissner, On variational theorem in elasticity, J. Math. and Physics, 29 (1950)

include the problems of the theory of asymmetric elasticity.

Let us consider now the following functional

$I \equiv I(\gamma_{ji}, \varkappa_{ji}, \varphi_i, \sigma_{ji}, \mu_{ji})$

$$I = \int_V \{ W_\varepsilon - X_i u_i - Y_i \varphi_i - \sigma_{ji}[\gamma_{ji} - (u_{i,j} - \epsilon_{kji}\varphi_k)] - \\ - \mu_{ji}(\varkappa_{ji} - \varphi_{i,j})\} dV - \int_{A_\sigma} (\hat{p}_i u_i + \hat{m}_i \varphi_i)\, dA - \\ - \int_{A_u} [p_i(u_i - \hat{u}_i) + m_i(\varphi_i - \hat{\varphi}_i)]\, dA \,, \tag{3.3.1}$$

where

$$W_\varepsilon = \mu \gamma_{(ij)}\gamma_{(ij)} + \alpha \gamma_{\langle ij\rangle}\gamma_{\langle ij\rangle} + \gamma \varkappa_{(ij)}\varkappa_{(ij)} + \\ + \varepsilon \varkappa_{\langle ij\rangle}\varkappa_{\langle ij\rangle} + \frac{\lambda}{2}\gamma_{kk}\gamma_{nn} + \frac{\beta}{2}\varkappa_{kk}\varkappa_{nn} \,. \tag{3.3.2}$$

Here $\hat{p}_i$ and $\hat{m}_i$ are forces and moments given on A_σ, $\hat{u}_i$, $\hat{\varphi}_i$ -components of the displacement vector $\underline{u}$ and of the rotation vector on $\underline{\varphi}$, respectively. Let us seek for the conditions necessary for I to be stationary. Equalling the first variation I to zero and taking into account that functions γ_{ji}, $\varkappa_{ji}$, u_i, φ_i, σ_{ji}, μ_{ji} show virtual increments inside the volume V, while the virtual increments of functions u_i, φ_i can be arbitrary on A_σ, and the virtual increments of functions p_i, m_i -arbitrary on A_u, we obtain

$$\delta I = 0 = \int_V \Big\{ \frac{\partial W_\varepsilon}{\partial \gamma_{ji}}\delta\gamma_{ji} + \frac{\partial W_\varepsilon}{\partial \varkappa_{ji}}\delta\varkappa_{ji} - X_i \delta u_i - Y_i \delta\varphi_i - \delta\sigma_{j,i}[\gamma_{ji} + \\ - (u_{i,j} - \epsilon_{ijk}\varphi_i)] - \sigma_{ji}[\delta\gamma_{ji} - (\delta u_{i,j} - \epsilon_{ijk}\varphi_i)] + \tag{3.3.3}$$

$$- \delta\mu_{ji}[\varkappa_{ji} - \varphi_{i,j}] - \mu_{ji}(\delta\varkappa_{ji} - \delta\varphi_{i,j})\}dV - \int_{A_\sigma}(\hat{p}_i\delta u_i +$$

$$+ \hat{m}_i\delta\varphi_i)dA - \int_{A_u}[(u_i - \hat{u}_i)\delta p_i + (\varphi_i - \hat{\varphi}_i)\delta m_i]\, dA .$$

Integrating by parts, making use of Gauss' transformation and arranging the results in groups, we obtain

$$\begin{aligned} &\int_V \Big\{\Big(\frac{\partial W_\varepsilon}{\partial \gamma_{ji}} - \sigma_{ji}\Big)\delta\gamma_{ji} + \Big(\frac{\partial W_\varepsilon}{\partial \varkappa_{ji}} - \mu_{ji}\Big)\delta\varkappa_{ji} - (X_i + \sigma_{ji,j})\delta u_i - \\ &- (\epsilon_{ijk}\sigma_{jk} + \mu_{ji,j} + Y_i)\delta\varphi_i + (\gamma_{ji} - u_{i,j} + \epsilon_{kji}\varphi_k)\delta\sigma_{ji} + \\ &+ (\varkappa_{ji} - \varphi_{i,j})\delta\mu_{ji}\Big\}dV - \int_{A_\sigma}[(p_i - \hat{p}_i)\delta u_i + (m_i - \hat{m}_i)\delta\varphi_i]dA - \\ &- \int_{A_u}[(u_i - \hat{u}_i)\delta p_i + (\varphi_i - \hat{\varphi}_i)\delta m_i]\, dA = 0 . \end{aligned} \tag{3.3.4}$$

As the result of independence of particular increments $\delta\gamma_{ji}$, $\delta\varkappa_{ji}$, δu_i, $\delta\varphi_i$, $\delta\sigma_{ji}$, $\delta\mu_{ji}$ from each other, we obtain from Eq. (3.3.4) the following system of Euler equations of variational problem

$$\begin{cases} \sigma_{ji,j} + X_i = 0 , & \epsilon_{ijk}\sigma_{jk} + \mu_{ji,j} + Y_i = 0 , \\ \gamma_{ji} = u_{i,j} - \epsilon_{ijk}\varphi_k , & \varkappa_{ji} = \varphi_{i,j} , \\ \dfrac{\partial W_\varepsilon}{\partial \gamma_{ji}} = \sigma_{ji} , \quad \dfrac{\partial W_\varepsilon}{\partial \varkappa_{ji}} = \mu_{ji} , & \underline{x} \in V , \end{cases} \tag{3.3.5}$$

$$\begin{cases} p_i = \hat{p}_i , \quad m_i = \hat{m}_i , & \underline{x} \in A_\sigma , \\ u_i = \hat{u}_i , \quad \varphi_i = \hat{\varphi}_i , & \underline{x} \in A_u . \end{cases} \tag{3.3.6}$$

This is the basic system of equations of the theory of asymmetric elasticity. The theorem of E. Reissner extended to the problems of asymmetric elasticity states that from among all the stress states σ_{ji}, couple-stresses states μ_{ji}, displacement states u_i and rotation states φ_i, satisfying the boundary conditions (3.3.6) and equilibrium equations (3.3.5) - only those actually appear which reduce the functional I to a minimum.

3. 4 Plane State of Strain.

In the plane state of strain all causes and effects depend on two variables only. Assuming that the displacements and rotations do not depend on the variable x_3, we have

$$\underline{u} = (u_1, u_2, 0) \,, \qquad \underline{\varphi} = (0, 0, \varphi_3) \,, \tag{3.4.1}$$

where u_1, u_2, φ_3 are functions of the variables x_1, x_2.

In accordance with the definition, we obtain for the plane state of strain the following components of the tensors γ_{ji} and $\varkappa_{ji}$:

$$\begin{aligned} &\gamma_{11} = \partial_1 u_1 \,, \quad \gamma_{22} = \partial_2 u_2 \,, \quad \gamma_{12} = \partial_1 u_2 - \varphi_3 \\ &\gamma_{21} = \partial_2 u_1 + \varphi_3 \,, \quad \varkappa_{13} = \partial_1 \varphi_3 \,, \quad \varkappa_{23} = \partial_2 \varphi_3 \,. \end{aligned} \tag{3.4.2}$$

The remaining values γ_{ji} and $\varkappa_{ji}$ are equal to zero. From the relations (3.4.2) we get

$$\sigma_{ji} = (\mu + \alpha)\gamma_{ji} + (\mu - \alpha)\gamma_{ij} + \lambda \gamma_{kk}\delta_{ji} \,, \tag{3.4.3}$$

$$\sigma_{33} = \lambda\gamma_{kk}\,,\ \mu_{j3} = (\gamma+\varepsilon)\varkappa_{j3}\,,\ \mu_{3j} = (\gamma-\varepsilon)\varkappa_{j3}\,,\ j = 1,2.$$

Here $\gamma_{kk} = \gamma_{11} + \gamma_{22}$. The state of stress σ_{ji} and the state of couple-stress μ_{ji} are characterized by the matrices

$$\underline{\sigma} = \begin{vmatrix} \sigma_{11} & \sigma_{12} & 0 \\ \sigma_{21} & \sigma_{22} & 0 \\ 0 & 0 & \sigma_{33} \end{vmatrix}\,,\quad \underline{\mu} = \begin{vmatrix} 0 & 0 & \mu_{13} \\ 0 & 0 & \mu_{23} \\ \mu_{31} & \mu_{32} & 0 \end{vmatrix}\,. \tag{3.4.4}$$

The equations of equilibrium (3.1.5) for the plane state of strain are reduced to three equations, namely

$$\begin{aligned} &\partial_1\sigma_{11} + \partial_2\sigma_{21} = 0\,, \\ &\partial_1\sigma_{12} + \partial_2\sigma_{22} = 0\,, \\ &\sigma_{12} - \sigma_{21} + \partial_1\mu_{13} + \partial_2\mu_{23} = 0\,. \end{aligned} \tag{3.4.5}$$

By eliminating the stresses from Eqs.(3.4.5) and taking into consideration Eqs.(3.4.2) and (3.4.3), we arrive at the following set of three equations:

$$\begin{aligned} &(\mu+\alpha)\nabla_1^2 u_1 + (\mu+\lambda-\alpha)\partial_1 e + 2\alpha\partial_2\varphi_3 = 0\,, \\ &(\mu+\alpha)\nabla_1^2 u_2 + (\mu+\lambda-\alpha)\partial_2 e - 2\alpha\partial_1\varphi_3 = 0\,, \\ &[(\gamma+\varepsilon)\nabla_1^2 - 4\alpha]\varphi_3 + 2\alpha(\partial_1 u_2 - \partial_2 u_1) = 0\,. \end{aligned} \tag{3.4.6}$$

We have

$$\partial_1 u_1 + \partial_2 u_2 = e\,, \qquad \nabla_1^2 = \partial_1^2 + \partial_2^2\,.$$

Let us return to the formulae (3.4.2). It is readily observed that the quantities appearing in these formulae are connected by means of the relations

$$\partial_1\gamma_{21} - \partial_2\gamma_{11} - \varkappa_{13} = 0 \ , \qquad \partial_1\gamma_{22} - \partial_2\gamma_{12} - \varkappa_{23} = 0 \ , \qquad (3.4.7)$$

$$\partial_1\varkappa_{23} - \partial_2\varkappa_{13} = 0 \ ,$$

which can also be written in the form

$$\partial_1^2\gamma_{22} + \partial_2^2\gamma_{11} = \partial_1\partial_2\,(\gamma_{12} + \gamma_{21}) \ ,$$

$$\partial_2^2\gamma_{12} - \partial_1^2\gamma_{21} = \partial_1\partial_2\,(\gamma_{22} - \gamma_{11}) - (\partial_1\varkappa_{13} - \partial_2\varkappa_{23}), \qquad (3.4.8)$$

$$\partial_1\varkappa_{23} - \partial_2\varkappa_{13} = 0 \ .$$

These are the compatibility equations for the two-dimensional problem of micropolar medium.

By solving Eqs.(3.1.3) for the strains γ_{ji} and $\varkappa_{j3}$ $(j = 1, 2)$, we have

$$\gamma_{11} = \frac{1}{2\mu}\left[\sigma_{11} - \frac{\lambda}{2(\lambda+\mu)}(\sigma_{11} + \sigma_{22})\right],$$

$$\gamma_{22} = \frac{1}{2\mu}\left[\sigma_{22} - \frac{\lambda}{2(\lambda+\mu)}(\sigma_{11} + \sigma_{22})\right],$$

$$\gamma_{12} = \frac{1}{4\mu}(\sigma_{12} + \sigma_{21}) + \frac{1}{4\alpha}(\sigma_{12} - \sigma_{21}) \ ,$$

$$\gamma_{21} = \frac{1}{4\mu}(\sigma_{12} + \sigma_{21}) + \frac{1}{4\alpha}(\sigma_{21} - \sigma_{12}) \ .$$

By introducing the above relations into the compatibility equations, we arrive at the following three equations in stresses:

(3.4.9)
$$\partial_2^2\sigma_{11}+\partial_1^2\sigma_{22}-\frac{\lambda}{2(\lambda+\mu)}\nabla_1^2(\sigma_{11}+\sigma_{22})=\partial_1\partial_2(\sigma_{12}+\sigma_{21}) \quad ,$$
$$(\partial_2^2-\partial_1^2)(\sigma_{12}+\sigma_{21})+\frac{\mu}{\alpha}\nabla_1^2(\sigma_{12}-\sigma_{21})=2\partial_1\partial_2(\sigma_{22}-\sigma_{11})-\frac{4\mu}{\gamma+\varepsilon}(\partial_1\mu_{13}+\partial_2\mu_{23}),$$
$$\partial_1\mu_{23}-\partial_2\mu_{13}=0 \ .$$

The above equations are deduced by H. Schaefer+).

We now introduce the stress functions F and Ψ and connect them with the stresses by the relations++)

(3.4.10)
$$\sigma_{11}=\partial_2^2F-\partial_1\partial_2\Psi, \qquad \sigma_{22}=\partial_1^2F+\partial_1\partial_2\Psi,$$
$$\sigma_{12}=-\partial_1\partial_2F-\partial_2^2\Psi, \qquad \sigma_{21}=-\partial_1\partial_2F+\partial_1^2\Psi,$$
$$\mu_{13}=\partial_1\Psi, \qquad \mu_{23}=\partial_2\Psi .$$

By substituting relations (3.4.10) into Eqs.(3.4.5), we find that they are identically satisfied. By substituting, in turn, (3.4.10) into the first and second compatibility equations (3.4.9), we obtain

(3.4.11)
$$\nabla_1^2\nabla_1^2F=0 ,$$
$$\nabla_1^2(l^2\nabla_1^2-1)\Psi=0 ,$$

+) H. Schaefer, Versuch einer Elastizitätstheorie der zweidimensionalen eben Cosserat-Kontinuums, Misz.Angew.Math.Festschrift Tollmien, Akademie Vlg.,Berlin 1962.

++)R.D. Mindlin, H.F. Tiersten, Effects of couple-stresses in linear elasticity, Arch.Rat.Mech.Anal., 11 (1962).

where

$$l^2 = \frac{(\gamma+\varepsilon)(\mu+\alpha)}{4\mu\alpha} .$$

The functions F and Ψ are not independent. They are connected by the first and the second relation (3.4.7). Consequently, we obtain

$$\begin{gathered} -\partial_1(1 - l^2\nabla_1^2)\Psi = A\,\partial_2\nabla_1^2 F, \\ \partial_2(1 - l^2\nabla_1^2)\Psi = A\,\partial_1\nabla_1^2 F, \\ A = \frac{(\lambda+2\mu)(\gamma+\varepsilon)}{4\mu(\lambda+\mu)} . \end{gathered} \tag{3.4.12}$$

We have still to give the boundary conditions for Eqs.(3.4.11).

The boundary conditions have the following form

$$\sigma_{ji} n_j = p_i \ , \quad \mu_{ji} n_j = m_i \ , \quad i,j = 1,2 . \tag{3.4.13}$$

These equations, expressed in Ψ and F , lead to the following ones

$$\begin{gathered} \frac{d}{ds}\left(\frac{\partial F}{\partial x_2} - \frac{\partial \Psi}{\partial x_1}\right) = p_1 \ , \qquad \frac{d}{ds}\left(\frac{\partial F}{\partial x_1} + \frac{\partial \Psi}{\partial x_2}\right) = -p_2 \ , \\ \frac{\partial \Psi}{\partial x_1} n_1 + \frac{\partial \Psi}{\partial x_2} n_2 = m_3 . \end{gathered} \tag{3.4.14}$$

Here

$$n_1 = \cos(\underline{n}, x_1) = \frac{dx_1}{dn} = \frac{dx_2}{ds} \ , \quad n_2 = \cos(\underline{n}, x_2) = \frac{dx_2}{dn} = -\frac{dx_1}{ds} .$$

The quantities $\frac{d}{ds}$ and $\frac{d}{dn}$ are the derivatives along the boundary s and along the normale to this boundary.

The boundary conditions (3.4.14) may be written in the form

$$\frac{\partial F}{\partial n} + \frac{\partial \Psi}{\partial s} = f_1 n_1 + f_2 n_2 , \quad \frac{\partial F}{\partial s} - \frac{\partial \Psi}{\partial n} = f_2 n_1 - f_1 n_2 , \quad \frac{\partial \Psi}{\partial n} = m_3 , \tag{3.4.15}$$

where

$$f_1 = -\int_{s_0}^{s} p_2(s)\,ds , \qquad f_2 = \int_{s_0}^{s} p_1(s)\,ds .$$

Let us consider the case of an elastic semi-space $x_1 \geq 0$, loaded on the boundary $x_1 = 0$ by forces $p(x_2)$. In this case the solution of Eqs.(3.4.11) can be represented in the form of Fourier integrals

$$\begin{aligned} F &= \frac{1}{\sqrt{2\pi}} \int_{-\infty}^{\infty} (M + N\xi x_1)\, e^{-\xi x_1 - i\xi x_2}\, d\xi , \\ \Psi &= \frac{1}{\sqrt{2\pi}} \int_{-\infty}^{\infty} (C e^{-\xi x_1} + D e^{-\sigma x_1}) e^{-i\xi x_2}\, d\xi , \qquad \sigma = \left(\xi + \frac{1}{l^2}\right)^{1/2} . \end{aligned} \tag{3.4.16}$$

The boundary conditions have the following form

$$\sigma_{11}(0, x_2) = -p(x_2), \; \sigma_{12}(0, x_2) = 0 , \; \mu_{13}(0, x_2) = 0 .$$

We shall express these conditions by means of the functions Ψ and F. However, we have four constants of integration C, D, M, N and only three boundary conditions (3.4.17). Besides we have Eqs.(3.4.12) connecting the functions Ψ and F. These relations, together with the boundary conditions (3.4.17), permit the unique determination of the constants of integration.

3. 5 Plane State of Stress.

Consider a cylinder with generators parallel to the x_3-axis and the bases in the planes $x_3 = \pm h$. This cylinder will be called if its height $2h$ is small as compared with the linear dimensions of the cross-section. Assume that the side surface of the cylinder is loaded by the forces $\underline{p}$ and moments $\underline{m}$, where

$$\underline{p} = (p_1, p_2, 0) \quad , \qquad \underline{m} = (0, 0, m_3) \qquad (3.5.1)$$

We further assume that the loadings p_1, p_2 and the moment m_3 are distributed symmetrically with respect to the middle plane $x_3 = 0$.

Suppose that the plate is also acted upon by the body forces $\underline{X}$ and the moments $\underline{Y}$, i.e.

$$\underline{X} = (X_1, X_2, 0) \quad , \qquad \underline{Y} = (0, 0, Y_3) \qquad (3.5.2)$$

also symmetric with respect to the middle plane.

Under the action of these loadings there exists in the plate a state of stress, in general spatial. There occur all components of the state of stress σ_{ji}, μ_{ji} as functions of x_1, x_2, x_3. We assume that the planes $x_3 = \pm h$ are free of stress, i.e.

$$\sigma_{33} = \sigma_{31} = \sigma_{32} = 0 \, , \; \mu_{33} = \mu_{31} = \mu_{32} = 0 \, , \text{ for } \; x_3 = \pm h . \qquad (3.5.3)$$

Consider the first three equations of equilibrium

$$
\begin{aligned}
&\partial_1\sigma_{11} + \partial_2\sigma_{21} + \partial_3\sigma_{31} + X_1 = 0 \,, \\
(3.5.4) \qquad &\partial_1\sigma_{12} + \partial_2\sigma_{22} + \partial_3\sigma_{32} + X_2 = 0 \,, \\
&\partial_1\sigma_{13} + \partial_2\sigma_{23} + \partial_3\sigma_{33} \qquad = 0 \,.
\end{aligned}
$$

In view of the symmetric distribution of the body forces X_1, X_2 and the forces p_1, p_2 with respect to the middle plane, the stresses σ_{11}, σ_{22}, σ_{12} and σ_{21} are symmetric and σ_{31}, σ_{32} antisymmetric with respect to this plane. The stress σ_{33} is a symmetric function of x_3 and therefore σ_{13} and σ_{23} are antisymmetric with respect to the variable x_3. On the basis of the remaining equilibrium equations

$$
\begin{aligned}
&\sigma_{23} - \sigma_{32} + \partial_1\mu_{11} + \partial_2\mu_{21} + \partial_3\mu_{31} = 0 \,, \\
(3.5.5) \qquad &\sigma_{31} - \sigma_{13} + \partial_1\mu_{12} + \partial_2\mu_{22} + \partial_3\mu_{32} = 0 \,, \\
&\sigma_{12} - \sigma_{21} + \partial_1\mu_{13} + \partial_2\mu_{23} + \partial_3\mu_{33} + Y_3 = 0 \,,
\end{aligned}
$$

we find that in view of the antisymmetry of the stresses σ_{23}, σ_{32}, σ_{31}, σ_{13} with respect to the plane $x_3 = 0$ the stresses μ_{11}, μ_{21}, μ_{12} and μ_{22} are antisymmetric and the stresses μ_{31}, μ_{32} symmetric with respect to this plane. In view of the symmetry of the stresses σ_{12}, σ_{21} and the symmetry of the body moment Y_3 with respect to the middle plane $x_3 = 0$, follows from the last equation (3.5.5) that μ_{13}, μ_{23} are symmetric and the stress μ_{33} an tisymmetric with respect to the plane $x_3 = 0$.

Let us integrate over the thickness of the first two equations of the group (3.5.4) and the last Eq.(3.5.5)

$$\int_{-h}^{h}(\partial_1\sigma_{11} + \partial_2\sigma_{21} + \partial_3\sigma_{31} + X_1)\,dx_3 = 0\,,$$
$$\int_{-h}^{h}(\partial_1\sigma_{12} + \partial_2\sigma_{22} + \partial_3\sigma_{23} + X_2)\,dx_3 = 0\,, \qquad (3.5.6)$$
$$\int_{-h}^{h}(\sigma_{12} - \sigma_{21} + \partial_1\mu_{13} + \partial_2\mu_{23} + \partial_3\mu_{33} + Y_3)\,dx_3 = 0\,.$$

Observe that

$$\int_{-h}^{h}\partial_3\sigma_{3j}\,dx_3 = \sigma_{3j}(x_1, x_2, \pm h) = 0\,, \quad j = 1, 2\,,$$
$$\int_{-h}^{h}\partial_3\mu_{33}\,dx_3 = \mu_{33}(x_1, x_2, \pm h) = 0\,.$$

The above integrals vanish in view of the boundary conditions (3.5.3). Equation (3.5.6) can be represented in the form

$$\partial_1\sigma_{11}^* + \partial_2\sigma_{21}^* + X_1^* = 0\,,$$
$$\partial_1\sigma_{12}^* + \partial_2\sigma_{22}^* + X_2^* = 0\,, \qquad (3.5.7)$$
$$\sigma_{12}^* - \sigma_{21}^* + \partial_1\mu_{13}^* + \partial_2\mu_{23}^* + Y_3^* = 0\,.$$

Here, the quantities

$$\sigma_{\nu\mu}^*(x_1, x_2) = \frac{1}{2h}\int_{-h}^{h}\sigma_{\nu\mu}(x_1,x_2,x_3)\,dx_3\,,\quad X_\nu^*(x_1,x_2) = \frac{1}{2h}\int_{-h}^{h}X_\nu(x_1,x_2,x_3)\,dx_3\,,$$

$$\mu^*_{\nu 3} = \frac{1}{2h}\int_{-h}^{h} \mu_{\nu 3}\, dx_3 \quad , \quad Y^*_3 = \frac{1}{2h}\int_{-h}^{h} Y_3\, dx_3 \quad , \quad \nu,\mu = 1,2 \ ,$$

are mean values of the stresses $\sigma_{\nu\mu}$, $\mu_{\nu 3}$, $(\nu,\mu = 1,2)$ the body forces X_ν and the body moments Y_3 along the thickness of the plate. We arrived at a system of three equilibrium equations, in which the mean values of the stresses depend on the variables x_1 and x_2 only.

Let us integrate the third Eq.(3.5.4) and the first Eqs.(3.5.5) along the thickness of the plate. We obtain

$$\begin{aligned} &\partial_1 \sigma^*_{13} + \partial_2 \sigma^*_{23} = 0 \ , \\ &\sigma^*_{23} - \sigma^*_{32} + \partial_1 \mu^*_{11} + \partial_2 \mu^*_{21} = 0 \ , \qquad (3.5.8) \\ &\sigma^*_{31} - \sigma^*_{32} + \partial_1 \mu^*_{12} + \partial_2 \mu^*_{21} = 0 \ . \end{aligned}$$

We have used here the boundary conditions (3.5.3), for

$$\int_{-h}^{h} \partial_3 \mu_{3j}\, dx_3 = \mu_{3j}(x_1, x_2, \pm h) = 0 \ , \ j = 1,2 \ ,$$

$$\int_{-h}^{h} \partial_3 \sigma_{33}\, dx_3 = \sigma_{33}(x_1, x_2, \pm h) = 0 \ .$$

Equations (3.5.8) are identically satisfied in view of the antisymmetry of the functions σ_{13}, σ_{31}, σ_{32}, σ_{23}, μ_{11}, μ_{21}, μ_{12}, μ_{22}. Thus

$$\sigma^*_{13} = \sigma^*_{23} = \sigma^*_{31} = \mu^*_{11} = \mu^*_{21} = \mu^*_{12} = \mu^*_{22} = 0 \ .$$

We do not make an appreciable error by assuming that the stresses σ_{13}, σ_{31}, σ_{32}, σ_{23}, σ_{33} and μ_{11}, μ_{22}, μ_{12}, μ_{21} are very small as compared with σ_{11}, σ_{22}, σ_{12}, σ_{21}, μ_{13}, μ_{31}, μ_{23}, μ_{32}. This assumption is the better, considering the smaller the height of the plate as compared with the other linear dimensions of the plate.

Thus, in the thin plate, the state of stress is approximately described by the tensors

$$\underline{\sigma}^* = \begin{vmatrix} \sigma_{11}^* & \sigma_{12}^* & 0 \\ \sigma_{21}^* & \sigma_{22}^* & 0 \\ 0 & 0 & 0 \end{vmatrix}, \quad \underline{\mu}^* = \begin{vmatrix} 0 & 0 & \mu_{13}^* \\ 0 & 0 & \mu_{23}^* \\ \mu_{31}^* & \mu_{32}^* & 0 \end{vmatrix}. \qquad (3.5.9)$$

This state of stress will be called the generalized plane state of stress of the Cosserat medium. The state of displacements and rotations in the plate can be described by the mean values of the vectors

$$\underline{u}^* \equiv (u_1^*, u_2^*, 0) \quad , \qquad \underline{\varphi}^* \equiv (0, 0, \varphi_3^*), \qquad (3.5.10)$$

where

$$u_j^*(x_1, x_2) = \frac{1}{2h}\int_{-h}^{h} u_j(x_1, x_2, x_3)\,dx_3 \quad ,$$

$$\varphi_j^*(x_1, x_2) = \frac{1}{2h}\int_{-h}^{h} \varphi_j(x_1, x_2, x_3)\,dx_3\,, \quad j = 1, 2, 3.$$

Assuming the symmetry of the functions p_1, p_2, p_3 and the functions X_1, X_2, Y_3 with respect to the middle plane $x_3 = 0$ the displacement u_3 and the rotations φ_1, φ_2 vanish on this plane and are antisymmetric with respect to it. Hence, $u_3^* = 0$, $\varphi_1^* = \varphi_2^* = 0$. Observe, however, that the quantity $\partial_3 u_3$ is symmetric with respect to this plane and the quantity

$$\gamma_{33}^* = \frac{1}{2h}\int_{-h}^{h} \partial_3 u_3 \, dx_3$$

is different from zero.

Thus, we obtain a state of strain of the plate, described by the tensors

(3.5.11) $$\underline{\gamma}^* = \begin{vmatrix} \gamma_{11}^* & \gamma_{12}^* & 0 \\ \gamma_{21}^* & \gamma_{22}^* & 0 \\ 0 & 0 & \gamma_{33}^* \end{vmatrix}, \quad \underline{\varkappa}^* = \begin{vmatrix} 0 & 0 & \varkappa_{13}^* \\ 0 & 0 & \varkappa_{23}^* \\ 0 & 0 & 0 \end{vmatrix}.$$

We, now, proceed to the constitutive relations. Averaging the stresses over the thickness $2h$ yields

(3.5.12) $$\sigma_{ji}^* = (\mu+\alpha)\gamma_{ji}^* + (\mu-\alpha)\gamma_{ij}^* + \lambda\gamma_{kk}^*\delta_{ji},$$

(3.5.13) $$\mu_{ji}^* = (\gamma+\varepsilon)\varkappa_{ji}^* + (\gamma-\varepsilon)\varkappa_{ij}^* + \beta\varkappa_{kk}\delta_{ji}, \quad i,j,k = 1,2,3.$$

Let us first determine the quantity γ_{33}^* from the condition $\sigma_{33}^* = 0$.

(3.5.14) $$\gamma_{33}^* = -\frac{\lambda}{\lambda+2\mu}(\gamma_{11}^* + \gamma_{22}^*).$$

By taking into account (3.5.14) and the matrices (3.5.9) and (3.5.11) we obtain the constitutive relations

$$\sigma_{12}^{*} = 2\mu\left[\gamma_{11}^{*} + \frac{\lambda}{\lambda+2\mu}(\gamma_{11}^{*} + \gamma_{22}^{*})\right] ,$$

$$\sigma_{22}^{*} = 2\mu\left[\gamma_{22}^{*} + \frac{\lambda}{\lambda+2\mu}(\gamma_{11}^{*} + \gamma_{22}^{*})\right] ,$$

$$\sigma_{12}^{*} = (\mu+\alpha)\gamma_{12}^{*} + (\mu-\alpha)\gamma_{21}^{*} , \quad \sigma_{21}^{*} = (\mu+\alpha)\gamma_{21}^{*} + (\mu-\alpha)\gamma_{12}^{*} , \tag{3.5.15}$$

$$\mu_{13}^{*} = (\gamma+\varepsilon)\varkappa_{13}^{*} , \qquad \mu_{23}^{*} = (\gamma+\varepsilon)\varkappa_{23}^{*} ,$$

$$\mu_{31}^{*} = (\gamma-\varepsilon)\varkappa_{13}^{*} , \qquad \mu_{32}^{*} = (\gamma-\varepsilon)\varkappa_{23}^{*} .$$

Let us introduce the relations (3.5.16) into the equilibrium equations (3.5.7), making use of the formulae

$$\gamma_{11}^{*} = \partial_1 u_1^{*} , \quad \gamma_{22}^{*} = \partial_2 u_2^{*} , \quad \gamma_{12}^{*} = \partial_1 u_2^{*} - \varphi_3^{*} , \tag{3.5.16}$$

$$\gamma_{21}^{*} = \partial_2 u_1^{*} + \varphi_3^{*} , \quad \varkappa_{13}^{*} = \partial_1 \varphi_3^{*} , \quad \varkappa_{23}^{*} = \partial_2 \varphi_3^{*} . \tag{3.5.17}$$

Then we arrive at the system of differential equations in displacements and rotations

$$\begin{aligned}
&(\mu+\alpha)\nabla_1^2 u_1 + \beta_0 \partial_1 e^{*} + 2\alpha\partial_2\varphi_3^{*} + X_1^{*} = 0 , \\
&(\mu+\alpha)\nabla_1^2 u_2^{*} + \beta_0 \partial_2 e^{*} - 2\alpha\partial_1\varphi_3^{*} + X_2^{*} = 0 , \\
&[(\gamma+\varepsilon)\nabla_1^2 - 4\alpha]\varphi_3^{*} + 2\alpha(\partial_1 u_2^{*} - \partial_2 u_1^{*}) + Y_3^{*} = 0 ,
\end{aligned} \tag{3.5.18}$$

where

$$e^{*} = \partial_1 u_1^{*} + \partial_2 u_2^{*} , \quad \beta_0 = \frac{\mu(3\lambda+2\mu) - \alpha(\lambda+2\mu)}{\lambda+2\mu} . \tag{3.5.19}$$

In the plane state of stress there exists also a representation of stresses by the functions F, Ψ.

Below we shall present a different procedure for deriving the differential equations for the functions F and Ψ, analogous to Eqs.(3.4.11).

Let us contract the first two Eqs.(3.5.18). We obtain

$$\frac{4\mu(\lambda+\mu)}{\lambda+2\mu}\nabla_1^2 e^* + \partial_1 X_1^* + \partial_2 X_2^* = 0 . \tag{3.5.20}$$

In the relation (3.5.15) we have

$$e^* = \frac{\lambda+2\mu}{2\mu(3\lambda+2\mu)}(\sigma_{11}^* + \sigma_{22}^*) . \tag{3.5.21}$$

By introducing this expression into (3.5.20) and making use of the representation by the functions F and Ψ,

$$\begin{aligned} &\sigma_{11}^* = \partial_2^2 F - \partial_1\partial_2\Psi , && \sigma_{22}^* = \partial_1^2 F + \partial_1\partial_2\Psi , \\ &\sigma_{12}^* = -\partial_1\partial_2 F - \partial_2^2\Psi , && \sigma_{21}^* = -\partial_1\partial_2 F + \partial_1^2\Psi , \\ &\mu_{13}^* = \partial_1\Psi , && \mu_{23}^* = \partial_2\Psi , \end{aligned} \tag{3.5.22}$$

we reduce Eq.(3.5.21) to the form

$$\nabla_1^2\nabla_1^2 F + \frac{3\lambda+2\mu}{2(\lambda+\mu)}(\partial_1 X_1^* + \partial_2 X_2^*) = 0 . \tag{3.5.23}$$

Let us now differentiate the second relation of Eq.(3.5.18) with respect to x_1 and the first relation of Eq.(3.5.18) with respect to x_2 and subtract the results.

Then

$$\nabla_1^2(\partial_1 u_2^* - \partial_2 u_1^*) = \frac{2\alpha}{\mu+\alpha}\nabla_1^2\varphi_3^* - \frac{1}{\mu+\alpha}(\partial_1 X_2^* - \partial_2 X_1^*) . \quad (3.5.24)$$

By applying the operator ∇_1^2 to the third relation of Eq.(3.5.18) and making use of Eq.(3.5.24), we obtain

$$\nabla^2(1 - l^2\nabla_1^2)\Psi = -\frac{1}{2\mu}(\partial_1 X_2^* - \partial_2 X_1^*) + \frac{1}{\gamma+\varepsilon} l^2\nabla_1^2 Y_3^* , \quad (3.5.25)$$

where

$$l^2 = \frac{(\mu+\alpha)(\gamma+\varepsilon)}{4\mu\alpha} .$$

The functions F and Ψ are not independent. The equations

$$\partial_1\gamma_{21}^* - \partial_2\gamma_{11}^* - \varkappa_{13}^* = 0 \quad , \qquad \partial_1\gamma_{22}^* - \partial_2\gamma_{12}^* - \varkappa_{23}^* = 0 , \quad (3.5.26)$$

following from the relations (3.5.17) yield after having expressed strains by stresses, the following relations connecting the functions F and Ψ :

$$\begin{aligned} -\partial_1(1 - l^2\nabla_1^2)\Psi &= A\,\partial_2\nabla_1^2 F , \\ \partial_2(1 - l^2\nabla_1^2)\Psi &= A\,\partial_1\nabla_1^2 F . \end{aligned} \quad (3.5.27)$$

Here

$$A = \frac{(\lambda+\mu)(\gamma+\varepsilon)}{\mu(3\lambda+2\mu)} .$$

The differential equations for the functions F, Ψ and relations of the type (3.5.27) differ in the two states (the plane state of strain and the plane state of stress) only in the values of the coefficients. It can also be easily proved that in the case of the plane state of stress we obtain the boundary conditions analogous to (3.4.8). Namely, we have

$$\frac{\partial F}{\partial n} + \frac{\partial \Psi}{\partial s} = f_1 n_1 + f_2 n_2 , \quad \frac{\partial F}{\partial s} - \frac{\partial \Psi}{\partial n} = f_2 n_1 - f_1 n_2 , \tag{3.5.28}$$
$$\frac{\partial \Psi}{\partial n} = m_3^* ,$$

where

$$f_1 = -\int_{s_0}^{s} p_2^* ds , \quad f_2 = \int_{s_0}^{s} p_1^* ds , \quad p_\beta^*(x_1, x_2) = \frac{1}{2h}\int_{-h}^{h} p_\beta(x_1, x_2, x_3) dx_3 , \; \beta = 1,2.$$

In the two-dimensional problems of the theory of elasticity an important role is played by the complex variable method. The method was introduced by G.V. Kolosov+) and was developed by I.N. Muskhelishvili++) and his school.

Below we shall outline the applications of the complex variable functions+++). The main points of the method will be presented on the example of the plane state of strain.

First of all, we note that the differential relations (3.5.27) can be represented in the following form

$$-\frac{\partial Q}{\partial x_1} = \frac{\partial P}{\partial x_2} , \qquad \frac{\partial Q}{\partial x_2} = \frac{\partial P}{\partial x_1} , \tag{3.5.29}$$

+) G.V. Kolosov: Comp. Rend. Acad. Sci. Paris, 146 (1908), 522, 62 (1914), 384.

++) N.I. Muskhelishvili: Math. Ann. 107 (1932), 282, and monograph "Some Basic Problems of the Mathematical Theory of Elasticity" (in Russian, Moscow 1934), an English translation of 1949 edition of the book, P.Noordhoff, N.V.,of Groningen, 1953.

+++) G.N. Savin: "The distribution of stresses around holes" (in Russian) Moscow, 1968.

where

$$P = A\nabla_1^2 F \quad , \qquad Q = (1 - l^2\nabla_1^2)\Psi \quad .$$

Relations (3.5.29) represent the Cauchy-Riemann relations for the harmonic functions P and Q. The fact that these functions are harmonic can be verified from the equations

$$\nabla_1^2\nabla_1^2 F = 0 \quad , \qquad Q = (1 - l^2\nabla_1^2)\Psi = 0 \, . \qquad (3.5.30)$$

The function F, satisfying the first relation of the bi-harmonic equation (3.5.30) can be represented, according to E.Goursat's theorem, by the complex potentials φ and χ +)

$$F = \frac{1}{2}\left(\bar{z}\,\varphi(z) + z\,\overline{\varphi(z)} + \chi(z) + \overline{\chi(z)}\right) \quad , \qquad (3.5.31)$$

or

$$F = \mathrm{Re}\left[\bar{z}\,\varphi(z) + \chi(z)\right] \quad . \qquad (3.5.32)$$

Here φ and χ are the analytical functions of the complex variable

$$z = x_1 + ix_2 \;, \quad \bar{z} = x_1 - ix_2 \, .$$

Let us pass from the variables (x_1 , x_2) to (z , $\bar{z}$). We obtain

$$\frac{\partial}{\partial x_1} = \frac{\partial}{\partial z} + \frac{\partial}{\partial \bar{z}} \,, \quad \frac{\partial}{\partial x_2} = i\left(\frac{\partial}{\partial z} - \frac{\partial}{\partial \bar{z}}\right), \quad \frac{\partial^2}{\partial x_1 \partial x_2} = i\left(\frac{\partial^2}{\partial z^2} - \frac{\partial^2}{\partial \bar{z}^2}\right),$$

$$\frac{\partial^2}{\partial x_1^2} = \frac{\partial^2}{\partial z^2} + 2\frac{\partial}{\partial z \partial \bar{z}} + \frac{\partial^2}{\partial \bar{z}^2} \,, \quad \frac{\partial^2}{\partial x_2^2} = -\frac{\partial}{\partial z^2} + 2\frac{\partial^2}{\partial z \partial \bar{z}} - \frac{\partial^2}{\partial \bar{z}^2} \;,$$

+) See pag. 188.

$$\nabla_1^2 = \frac{\partial^2}{\partial x_1^2} + \frac{\partial^2}{\partial x_2^2} = 4\,\frac{\partial^2 z}{\partial z\,\partial \bar{z}} \quad .$$

Relations (3.5.29) in the variables (z , $\bar{z}$) can be written down in the form

$$(3.5.33) \qquad \begin{aligned} &\Big(\frac{\partial}{\partial z} + \frac{\partial}{\partial \bar{z}}\Big) Q(z,\bar{z}) = -i\Big(\frac{\partial}{\partial z} - \frac{\partial}{\partial \bar{z}}\Big) P(z,\bar{z}) ,\\ &i\Big(\frac{\partial}{\partial z} - \frac{\partial}{\partial \bar{z}}\Big) Q(z,\bar{z}) = \Big(\frac{\partial}{\partial z} + \frac{\partial}{\partial \bar{z}}\Big) P(z,\bar{z}) . \end{aligned}$$

Upon simplification we obtain

$$(3.5.34) \qquad \begin{aligned} &\frac{\partial}{\partial z} Q(z,\bar{z}) = -i\,\frac{\partial}{\partial z} P(z,\bar{z}) ,\\ &\frac{\partial}{\partial z} Q(z,\bar{z}) = i\,\frac{\partial}{\partial \bar{z}} P(z,\bar{z}) , \end{aligned}$$

where

$$Q(z,\bar{z}) = \Big(1 - 4l^2 \frac{\partial^2}{\partial z\,\partial \bar{z}}\Big)\Psi(z,\bar{z}) \;, \qquad P(z,\bar{z}) = 4A\,\frac{\partial^2 F(z,\bar{z})}{\partial z\,\partial \bar{z}} \quad .$$

Taking into account (3.5.32), relations (3.5.34) can be represented in the following form

$$(3.5.35) \qquad \begin{aligned} &\frac{\partial}{\partial z}\Big(1 - 4l^2 \frac{\partial}{\partial z\,\partial \bar{z}}\Big)\Psi(z,\bar{z}) = -2i\varphi''(z) ,\\ &\frac{\partial}{\partial \bar{z}}\Big(1 - 4l^2 \frac{\partial}{\partial z\,\partial \bar{z}}\Big)\Psi(z,\bar{z}) = 2i\overline{\varphi''(z)} . \end{aligned}$$

If the potential φ is known, we can determine, from Eqs.(3.5.35), by the integration of the function $(1 - 4l^2 \frac{\partial}{\partial z\,\partial \bar{z}})\Psi(z,\bar{z})$. Thus, formulae (3.5.32) and (3.5.35) give the representation of the function $F(z,\bar{z})$ and $(1 - 4l^2 \frac{\partial}{\partial z\,\partial \bar{z}})\Psi(z,\bar{z})$ by the Kolosov--Muskhelishvili potentials.

The first and second differential equations (3.5.30) expressed in terms of the variables $(z, \bar{z})$ take the following form

$$\frac{\partial^4 F}{\partial z^2 \partial \bar{z}^2} = 0 \; , \qquad \frac{\partial^2}{\partial z \partial \bar{z}}\left(1 - 4l^2 \frac{\partial^2}{\partial z \partial \bar{z}}\right)\Psi(z, \bar{z}). \tag{3.5.36}$$

From Eq.(3.5.36) results that

$$\left(1 - 4l^2 \frac{\partial^2}{\partial z \partial \bar{z}}\right)\Psi(z, \bar{z}) = z f_1(\bar{z}) + f_2(z) \; , \tag{3.5.37}$$

where f_1 and f_2 are the analytical functions of their arguments. The boundary conditions of the problem can be represented in the following form, valid for a simply-connected region

$$\frac{\partial F}{\partial x_2} - \frac{\partial \Psi}{\partial x_1} = \int_s p_1 \, ds + C_1 \; ,$$

$$\frac{\partial F}{\partial x_1} + \frac{\partial \Psi}{\partial x_2} = -\int_s p_2 \, ds + C_2 \; , \qquad \int_s \frac{\partial \Psi}{\partial n} ds = \int_s m_3 \, ds + C_3 \, . \tag{3.5.38}$$

It will be more convenient to write down the first two boundary conditions in a complex form

$$\frac{\partial}{\partial z}(F + i\Psi) = -\frac{1}{2}\int_s (p_2 + ip_1) ds + \frac{1}{2}C_2 - \frac{1}{2}C_1 \, .$$

The third condition (3.5.38) remains unchanged.

If the function F and Ψ are already known, we can determine the stresses. Let us note that

$$\sigma_{11} + \sigma_{22} = 4\frac{\partial^2 F(z, \bar{z})}{\partial z \partial \bar{z}} \; ,$$

$$\sigma_{12} - \sigma_{21} = -4\,\frac{\partial^2\Psi(z,\bar{z})}{\partial z\,\partial\bar{z}} .$$

G.N. Savin+) and his collaborators solved a number of particular problems, especially concerning the concentration of stresses in the neighborhood of holes, by means of conformal mapping of a considered region on the unit circle.

3. 6 Galerkin's and Love's Generalized Functions.

Similarly as in elastokinetics, also, here the system of differential equations

$$(3.6.1) \qquad (\mu+\alpha)\nabla^2\underline{u} + (\lambda+\mu-\alpha)\,\mathrm{grad\,div}\,\underline{u} + 2\alpha\,\mathrm{rot}\,\underline{\varphi} + \underline{X} = 0 ,$$

$$(3.6.2) \qquad [(\gamma+\varepsilon)\nabla^2 - 4\alpha]\varphi + (\beta+\gamma-\varepsilon)\,\mathrm{grad\,div}\,\varphi + 2\alpha\,\mathrm{rot}\,u + Y = 0 ,$$

can be reduced to an explicit form by means of six solving functions $\underline{F}$, $\underline{M}$. By the reduction to an explicit form, we understand an operation which reduces the problem to solving such six equations so that each of them contains a single function only. Let us express the displacement $\underline{u}$ and rotation $\underline{\varphi}$ by means of two vector functions

$$\underline{u} = (\lambda+2\mu)\nabla^2[(\gamma+\varepsilon)\nabla^2 - 4\alpha]\underline{F} - [(\gamma+\varepsilon)(\lambda+\mu-\alpha)\nabla^2 +$$

$$(3.6.3) \qquad - 4\alpha(\lambda+\mu)]\,\mathrm{grad\,div}\,\underline{F} - 2\alpha[(\beta+2\gamma)\nabla^2 - 4\alpha]\,\mathrm{rot}\,\underline{M} ,$$

+) Savin, G.N.: See pag.188.

$$\underline{\varphi} = (\mu+\alpha)\nabla^2[(\beta+2\gamma)\nabla^2 - 4\alpha]\underline{M} - [(\mu+\alpha)(\beta+\gamma-\varepsilon)\nabla^2 +$$

$$- 4\alpha^2]\,\text{grad}\,\text{div}\,\underline{M} - 2\alpha(\lambda+2\mu)\nabla^2\text{rot}\,\underline{F} \,. \qquad (3.6.4)$$

These relations were obtained from the representation for the dynamical problem (see formulae (1.7.22) and (1.7.23) disregarding the dynamical terms. The material constants μ, λ, α, β, γ, ε, occurring in Eqs.(3.6.3) and (3.6.4) refer to an isothermal process.

Substituting (3.6.3) and (3.6.4) in the differential equations (3.6.1) and (3.6.2) we obtain the following differential equations for the vector functions $\underline{F}$ and $\underline{M}$:

$$(\lambda+2\mu)\nabla^2\nabla^2[(\mu+\alpha)(\gamma+\varepsilon)\nabla^2 - 4\alpha\mu]\,\underline{F} + \underline{X} = 0 \,, \qquad (3.6.5)$$

$$[(\beta+2\gamma)\nabla^2 - 4\alpha]\nabla^2[(\mu+\alpha)(\gamma+\varepsilon)\nabla^2 - 4\alpha]\,\underline{M} + \underline{Y} = 0 \,. \qquad (3.6.6)$$

The functions $\underline{F}$ and $\underline{M}$ should be regarded as Galerkin's vectors generalized on the micropolar continuum. If in relation (3.6.3) we set $\alpha = 0$, then we obtain the known representation of displacement $\underline{u}$ by the Galerkin vector $\underline{\Gamma}$ of the classical theory of elasticity

$$\underline{u} = \frac{\lambda+2\mu}{\mu}\left[\nabla^2\underline{\Gamma} - \frac{\lambda+\mu}{\lambda+2\mu}\,\text{grad}\,\text{div}\,\underline{\Gamma}\right] ,$$

$$\Gamma = \frac{\gamma+\varepsilon}{\mu}\,\nabla^2\underline{F} \,. \qquad (3.6.7)$$

Eq.(3.6.5) takes the following form

$$\nabla^2\nabla^2 \underline{\Gamma} + \frac{\underline{X}}{\lambda+2\mu} = 0 . \tag{3.6.8}$$

If we assume $\alpha=0$ in relation (3.6.4) and Eq.(3.6.6), we obtain an analogous relation for a body with the rotations only. Note that for an unbounded micropolar body the assumption $\underline{X}=0$ entails also $\underline{F}=0$. Similarly, for $\underline{Y}=0$ we have $\underline{M}=0$.

Eqs.(3.6.5) and (3.6.6) allow us to determine, in a very simple way, the fundamental solutions for an infinite micropolar space. In the classical theory of elasticity, when solving the axially symmetric problems, an important role is played by Love's function +). Below we shall derive two new functions constituting a generalization of Love's function on the case of micropolar elasticity.

Passing with Eqs.(3.6.1) and (3.6.2) to the (r,ϑ,z) coordinate system and assuming that both the causes and effects are independent of the angle ϑ, we obtain from Eqs.(3.6.1) and (3.6.2) two systems of equations independent of each other:

$$(\mu+\alpha)\left(\nabla^2-\frac{1}{r^2}\right)u_r+(\lambda+\mu-\alpha)\frac{\partial e}{\partial r}-2\alpha\frac{\partial\varphi_\vartheta}{\partial z}+X_r=0 ,$$

$$(\gamma+\varepsilon)\left(\nabla^2-\frac{1}{r^2}\right)\varphi_\vartheta-4\alpha\varphi_\vartheta+2\alpha\left(\frac{\partial u_r}{\partial z}-\frac{\partial u_z}{\partial r}\right)+Y_\vartheta=0 ,$$

+) Love, A.E.H., A treatise on the mathematical theory of elasticity. Cambridge, 1927, 4th edition, p.274.

$$(\mu+\alpha)\nabla^2 u_z+(\lambda+\mu-\alpha)\frac{\partial e}{\partial z}+2\alpha\frac{1}{r}\frac{\partial}{\partial r}(r\varphi_\theta)+X_z=0, \tag{3.6.9}$$

and

$$\begin{aligned}
&(\gamma+\varepsilon)\left(\nabla^2-\frac{1}{r^2}\right)\varphi_r-4\alpha\varphi_r+(\beta+\gamma-\varepsilon)\frac{\partial \varkappa}{\partial r}-2\alpha\frac{\partial u_\theta}{\partial z}+Y_r=0,\\
&(\mu+\alpha)\left(\nabla^2-\frac{1}{r^2}\right)u_\theta+2\alpha\left(\frac{\partial\varphi_r}{\partial z}-\frac{\partial\varphi_z}{\partial r}\right)+X_\theta=0,\\
&(\gamma+\varepsilon)\nabla^2\varphi_z-4\alpha\varphi_z+(\beta+\gamma-\varepsilon)\frac{\partial\varkappa}{\partial z}+\frac{2\alpha}{r}\frac{\partial}{\partial r}(r u_\theta)+Y_z=0.
\end{aligned} \tag{3.6.10}$$

In the above-mentioned equations the following notations have been introduced

$$\underline{u}=(u_r,u_\theta,u_z)\ ,\quad \underline{\varphi}=(\varphi_r,\varphi_\theta,\varphi_z)\ ,\quad \underline{X}=(X_r,X_\theta,X_z)\ ,$$

$$\underline{Y}=(Y_r,Y_\theta,Y_z)\ ,\quad e=\frac{1}{r}\frac{\partial}{\partial r}(u_r r)+\frac{\partial u_z}{\partial z}\ ,$$

$$\varkappa=\frac{1}{r}\frac{\partial}{\partial r}(r\varphi_r)+\frac{\partial\varphi_z}{\partial z}\ .$$

It is seen that the state of displacements and rotations $\underline{u}=(u_r,0,u_z)$, $\underline{\varphi}=(0,\varphi_\theta,0)$ appearing in the system of Eqs.(3.6.9) may be induced by the action of body forces and body couples in the form:

$$\underline{X}=(X_r,0,X_z)\ ,\quad \underline{Y}=(0,Y_\theta,0)$$

with boundary conditions chosen appropriately. Let us quote, as an example, that such a state may be induced by the action of a load parallel to the z-axis (e.g. Boussinesq problem for an elastic half-space).

The state of displacements and rotations

$$\underline{u} = (0, u_\theta, 0) \quad , \quad \underline{\varphi} = (\varphi_r, 0, \varphi_z) \quad ,$$

appearing in the system of Eqs.(3.6.10) may be due to the action of body forces and body couples

$$\underline{X} = (0, X_\theta, 0) \quad , \quad \underline{Y} = (Y_r, 0, Y_z) \quad ,$$

the boundary conditions being chosen appropriately, e.g., such a state of displacements and rotations will appear in the elastic half-space $z \geq 0$ loaded on the surface $z = 0$, by the twist moments with vectors directed along the z-axis.

Let us consider the system of Eqs.(3.6.9) under the assumption that $\underline{X} = (0, 0, X_z)$. Thereafter, we put into Eqs.(3.6.9) the quantities

$$(3.6.11) \qquad u_r = \frac{\partial v}{\partial r} \quad , \quad \varphi_\theta = \frac{\partial \omega}{\partial r} \quad , \quad u_z = w \; .$$

We take into account that

$$\left(\nabla^2 - \frac{1}{r^2}\right)\frac{\partial v}{\partial r} = \frac{\partial}{\partial r}\nabla^2 v \quad , \quad \left(\nabla^2 - \frac{1}{r^2}\right)\frac{\partial w}{\partial r} = \frac{\partial}{\partial r}\nabla^2 w .$$

By introducing the above expressions into the first and second Eqs.(3.6.9) we obtain a system of equations, wherein only the operators $\nabla^2 v, \nabla^2 \omega, \nabla^2 w$ appear. By integrating these equations with respect to r and taking into consideration the third Eq.(3.6.9), we get the following system of equations:

$$\left[(\lambda+2\mu)\nabla^2-(\lambda+\mu-\alpha)\frac{\partial^2}{\partial z^2}\right]v-2\alpha\frac{\partial\omega}{\partial z}+$$

$$+(\lambda+\mu-\alpha)\frac{\partial w}{\partial z}=0\,,$$

$$2\alpha\frac{\partial v}{\partial z}+\left[(\gamma+\varepsilon)\nabla^2-4\alpha\right]\omega-2\alpha w=0\,, \tag{3.6.12}$$

$$(\lambda+\mu-\alpha)\frac{\partial}{\partial z}\left(\nabla^2-\frac{\partial^2}{\partial z^2}\right)v+2\alpha\left(\nabla^2-\frac{\partial^2}{\partial z^2}\right)\omega+$$

$$+\left[(\mu+\alpha)\nabla^2+(\lambda+\mu-\alpha)\frac{\partial^2}{\partial z^2}\right]w+X_z=0\,.$$

This system of equations may be written also in the form

$$\begin{aligned} L_{vv}v+L_{v\omega}\omega+L_{vw}w&=0\,,\\ L_{\omega v}v+L_{\omega\omega}\omega+L_{\omega w}w&=0\,,\\ L_{wv}v+L_{w\omega}\omega+L_{ww}w&=0\,.\end{aligned} \tag{3.6.13}$$

Let us, now, introduce stress function, $\chi(r,z)$ connected with the functions v , ω , w by the following relations

$$v=\begin{vmatrix}0 & L_{v\omega} & L_{vw}\\ 0 & L_{\omega\omega} & L_{\omega w}\\ \chi & L_{w\omega} & L_{ww}\end{vmatrix},\quad \omega=\begin{vmatrix}L_{vv} & 0 & L_{vw}\\ L_{\omega v} & 0 & L_{\omega w}\\ L_{wv} & \chi & L_{ww}\end{vmatrix},\quad w=\begin{vmatrix}L_{vv} & L_{v\omega} & 0\\ L_{\omega v} & L_{\omega\omega} & 0\\ L_{wv} & L_{w\omega} & \chi\end{vmatrix} \tag{3.6.14}$$

wherefrom, we find that

$$u_r=\frac{\partial v}{\partial r}=-\frac{\partial}{\partial r\partial z}(\Gamma\chi)\,, \tag{3.6.15}$$

$$\varphi_\theta = \frac{\partial \omega}{\partial r} = 2\alpha(\lambda + 2\mu)\frac{\partial}{\partial r}(\nabla^2\chi) , \tag{3.6.16}$$

$$u_z = w = \Theta\chi - \frac{\partial^2}{\partial z^2}(\Gamma\chi) , \tag{3.6.17}$$

where

$$\Gamma = (\gamma + \varepsilon)(\lambda + \mu - \alpha)\nabla^2 - 4\alpha(\lambda + \mu) ,$$
$$\Theta = (\lambda + 2\mu)\nabla^2[(\gamma + \varepsilon)\nabla^2 - 4\alpha] .$$

By substituting (3.6.15)$\div$(3.6.17) into the last of Eqs.(3.6.13), we obtain the equation

$$(\lambda + 2\mu)\nabla^2\nabla^2[(\mu + \alpha)(\gamma + \varepsilon)\nabla^2 - 4\mu\alpha]\chi(r, z) + X_z(r, z) = 0. \tag{3.6.18}$$

This is the "sought for" generalization of Love's equation known from the classical theory of elasticity.

The state of displacements $\underline{u} \equiv (u_r, 0, u_z)$ and rotations $\underline{\varphi} \equiv (0, \varphi_\theta, 0)$, here considered, is connected with the state of stress by the following relations

$$\underline{\sigma} = \begin{vmatrix} \sigma_{rr} & 0 & \sigma_{rz} \\ 0 & \sigma_{\theta\theta} & 0 \\ \sigma_{zr} & 0 & \sigma_{zz} \end{vmatrix} , \quad \underline{\mu} = \begin{vmatrix} 0 & \mu_{r\theta} & 0 \\ \mu_{\theta r} & 0 & \mu_{\theta z} \\ 0 & \mu_{z\theta} & 0 \end{vmatrix} , \tag{3.6.19}$$

where

$$\begin{aligned} &\sigma_{rr} = 2\mu\frac{\partial u_r}{\partial r} + \lambda e , \quad \sigma_{\theta\theta} = 2\mu\frac{u_r}{r} + \lambda e , \\ &\sigma_{zz} = 2\mu\frac{\partial u_z}{\partial z} + \lambda e , \end{aligned} \tag{3.6.20}$$

$$\sigma_{rz} = \mu\left(\frac{\partial u_z}{\partial r} + \frac{\partial u_r}{\partial z}\right) - \alpha\left(\frac{\partial u_r}{\partial z} - \frac{\partial u_z}{\partial r}\right) + 2\alpha\varphi_\theta \ ,$$

$$\sigma_{zr} = \mu\left(\frac{\partial u_z}{\partial r} + \frac{\partial u_r}{\partial z}\right) + \alpha\left(\frac{\partial u_r}{\partial z} - \frac{\partial u_z}{\partial r}\right) - 2\alpha\varphi_\theta \ ,$$

and

$$u_{r\theta} = \gamma\left(\frac{\partial\varphi_\theta}{\partial r} - \frac{\varphi_\theta}{r}\right) + \varepsilon\left(\frac{\partial\varphi_\theta}{\partial r} + \frac{\varphi_\theta}{r}\right) \ ,$$

$$u_{\theta r} = \gamma\left(\frac{\partial\varphi_\theta}{\partial r} - \frac{\varphi_\theta}{r}\right) - \varepsilon\left(\frac{\partial\varphi_\theta}{\partial r} + \frac{\varphi_\theta}{r}\right) \ , \tag{3.6.21}$$

$$\mu_{\theta z} = (\gamma - \varepsilon)\frac{\partial\varphi_\theta}{\partial z} \ , \quad \mu_{z\theta} = (\gamma + \varepsilon)\frac{\partial\varphi_\theta}{\partial z} \ .$$

By expressing the above stresses by u_r , φ_θ , u_z from the formulae (3.6.15)÷(3.6.17), we get finally

$$\sigma_{rr} = -\frac{\partial}{\partial z}\left[2\mu\frac{\partial}{\partial r^2}(\Gamma\chi) - \lambda(\Lambda\chi)\right] \ ,$$

$$\sigma_{\theta\theta} = -\frac{\partial}{\partial z}\left[2\mu\frac{1}{r}\frac{\partial}{\partial r}(\Gamma\chi) - \lambda(\Lambda\chi)\right] \ , \tag{3.6.22}$$

$$\sigma_{zz} = -\frac{\partial}{\partial z}\left[2\mu\left(\frac{\partial^2\Gamma}{\partial z^2} - \Theta\right) - \lambda(\Lambda\chi)\right] \ ,$$

where

$$\Lambda = \nabla^2[(\mu+\alpha)(\gamma+\varepsilon)\nabla^2 - 4\alpha\mu] \ ,$$

$$\sigma_{rz} = \frac{\partial}{\partial r}\left[(\mu+\alpha)\Theta\chi - 2\mu\frac{\partial^2}{\partial z^2}(\Gamma\chi) + 4\alpha^2(\lambda+2\mu)\nabla^2\chi\right] \ , \tag{3.6.23}$$

$$\sigma_{zr} = \frac{\partial}{\partial r}\left[(\mu-\alpha)\Theta\chi - 2\mu\frac{\partial^2}{\partial z^2}(\Gamma\chi) - 4\alpha^2(\lambda+2\mu)\nabla^2\chi\right] \ ,$$

and

$$
(3.6.24)\qquad
\begin{aligned}
\mu_{r\theta} &= 2\alpha(\lambda+2\mu)\left[(\gamma+\varepsilon)\frac{\partial^2}{\partial r^2}-(\gamma-\varepsilon)\frac{1}{r}\frac{\partial}{\partial r}\right]\nabla^2\chi ,\\
\mu_{\theta r} &= 2\alpha(\lambda+2\mu)\left[(\gamma-\varepsilon)\frac{\partial^2}{\partial r^2}-(\gamma+\varepsilon)\frac{1}{r}\frac{\partial}{\partial r}\right]\nabla^2\chi ,\\
\mu_{\theta z} &= 2\alpha(\lambda+2\mu)(\gamma-\varepsilon)\frac{\partial^2}{\partial r\partial z}(\nabla^2\chi) ,\\
\mu_{z\theta} &= 2\alpha(\lambda+2\mu)(\gamma+\varepsilon)\frac{\partial^2}{\partial r\partial z}(\nabla^2\chi) .
\end{aligned}
$$

Now, let us pass to the classical theory of elasticity putting $\alpha = 0$ in (3.6.15)÷(3.6.24). Introducing, moreover, a new function

$$
(3.6.25)\qquad L(r,z) = (\gamma+\varepsilon)\nabla^2\chi(r,z)
$$

we obtain from (3.6.15)÷(3.6.17):

$$
(3.6.26)\qquad
\begin{aligned}
u_r &= -(\lambda+\mu)\frac{\partial^2 L}{\partial r\partial z} , \qquad \varphi_\theta = 0 ,\\
u_z &= (\lambda+2\mu)\nabla^2 L-(\lambda+\mu)\frac{\partial^2 L}{\partial z^2} .
\end{aligned}
$$

Eq.(3.6.18) transforms into

$$
(3.6.27)\qquad \nabla^2\nabla^2 L+\frac{1}{\mu(\lambda+2\mu)}X_z = 0 ,
$$

and - for $X_z = 0$ - becomes a bi-harmonic equation. The following formulae will describe the stresses:

$$
\begin{aligned}
\sigma_{rr}/\mu &= \frac{\partial}{\partial z}\left[\lambda\nabla^2 L-2(\lambda+\mu)\frac{\partial^2 L}{\partial r^2}\right] ,\\
\sigma_{\theta\theta}/\mu &= \frac{\partial}{\partial z}\left[\lambda\nabla^2 L-\frac{2}{r}(\lambda+\mu)\frac{\partial^2 L}{\partial z^2}\right] ,
\end{aligned}
$$

$$\sigma_{zz}/\mu = \frac{\partial}{\partial z}\left[(3\lambda+4\mu)\nabla^2 L - 2(\lambda+\mu)\frac{\partial^2 L}{\partial z^2}\right],$$

$$\sigma_{rz}/\mu = \frac{\partial}{\partial r}\left[(\lambda+2\mu)\nabla^2 L - 2(\lambda+\mu)\frac{\partial^2 L}{\partial z^2}\right], \qquad (3.6.28)$$

$$\mu_{r\theta} = \mu_{\theta r} = 0 \quad , \qquad \mu_{zr} = \mu_{rz} = 0 \quad .$$

In this way we arrived at the Love's stress function $L(r,z)$, known from the classical theory of elasticity+)

By returning now to the micropolar elasticity we shall present, (through example), the application of the stress function $\chi(r,z)$.

Consider an elastic half-space $z \geq 0$ subjected to the action of a loading $p(r)$ in parallel to the z-axis in the plane $z = 0$. The solution of Eq.(3.6.18) may be written in the form of Hankel's integral

$$\chi(r,z) = \int_0^\infty Z(z) J_0(\xi r)\, d\xi \qquad (3.6.29)$$

where

$$Z(z) = (A + B\xi)e^{-\xi z} + C e^{-\nu z} ,$$

$$\nu = (\xi^2 + \sigma^2)^{1/2} \quad , \qquad \sigma^2 = \frac{4\alpha\mu}{(\mu+\alpha)(\gamma+\varepsilon)} .$$

The function $Z(z)$ is chosen so as to have $\chi(r,z) \to 0$ for $z \to \infty$. The quantities A, B, C, functions of the parameter ξ, will be

+) Love A.E.H.: see pag. 191.

determined from the following three boundary conditions

$$(3.6.30)\quad \sigma_{zz}(r,0) = -p(r)\ ,\quad \sigma_{zr}(r,0) = 0\ ,\quad \mu_{z\vartheta}(r,0) = 0\ .$$

Here we make use of the formulas (3.6.22) ÷ (3.6.24), wherein the stresses σ_{zz}, σ_{zr}, $\mu_{z\vartheta}$ are expressed by the function χ .

We shall, now, consider the system of Eqs.(3.6.10), putting $\underline{Y} \equiv (0,0,Y_z)$. By substituting into these equations the following quantities

$$\varphi_r = \frac{\partial \eta}{\partial r}\ ,\qquad u_\vartheta = \frac{\partial \vartheta}{\partial r}\ ,\qquad \varphi_z = \tau$$

and bearing in mind

$$\left(\nabla^2 - \frac{1}{r^2}\right)\frac{\partial \eta}{\partial r} = \frac{\partial}{\partial r}\nabla^2 \eta\ ,\quad \left(\nabla^2 - \frac{1}{r^2}\right)\frac{\partial \vartheta}{\partial r} = \frac{\partial}{\partial r}\nabla^2 \vartheta\ ,$$

we obtain a system of equations, wherein only the Laplacians $\nabla^2\eta$, $\nabla^2\vartheta$ and $\nabla^2\tau$ occur. By integrating the first and second relations of Eqs.(3.6.10) with respect to r , we arrive at the following system of equations

$$\left[(\beta+2\gamma)\nabla^2 - 4\alpha - (\beta+\gamma-\varepsilon)\frac{\partial^2}{\partial z^2}\right]\eta - 2\alpha\frac{\partial \vartheta}{\partial z} + (\beta+\gamma-\varepsilon)\frac{\partial \tau}{\partial z} = 0\ ,$$

$$(3.6.31)\quad 2\alpha\frac{\partial \eta}{\partial z} + (\mu+\alpha)\nabla^2\vartheta - 2\alpha\tau = 0\ ,$$

$$(\beta+\gamma-\varepsilon)\left(\nabla^2 - \frac{\partial^2}{\partial z^2}\right)\frac{\partial \eta}{\partial z} + 2\alpha\left(\nabla^2 - \frac{\partial^2}{\partial z^2}\right)\vartheta + \left[(\gamma+\varepsilon)\nabla^2 - 4\alpha + (\gamma+\beta-\varepsilon)\frac{\partial^2}{\partial z^2}\right]\tau + Y_z = 0\ .$$

We may write this system in an abbreviated form

$$\begin{aligned} L_{\eta\eta}\eta + L_{\eta\vartheta}\vartheta + L_{\eta\tau}\tau &= 0 , \\ L_{\vartheta\eta}\eta + L_{\vartheta\vartheta}\vartheta + L_{\vartheta\tau}\tau &= 0 , \\ L_{\tau\eta}\eta + L_{\tau\vartheta}\vartheta + L_{\tau\tau}\tau &= 0 . \end{aligned} \tag{3.6.32}$$

Let us introduce the stress function $\Psi(r,z)$ connected with the functions η, ϑ, τ by the relations

$$\eta = \begin{vmatrix} 0 & L_{\eta\vartheta} & L_{\eta\tau} \\ 0 & L_{\vartheta\vartheta} & L_{\vartheta\tau} \\ \Psi & L_{\tau\vartheta} & L_{\tau\tau} \end{vmatrix} , \quad \vartheta = \begin{vmatrix} L_{\eta\eta} & 0 & L_{\eta\tau} \\ L_{\vartheta\eta} & 0 & L_{\vartheta\tau} \\ L_{\tau\eta} & \Psi & L_{\tau\tau} \end{vmatrix} , \quad \tau = \begin{vmatrix} L_{\eta\eta} & L_{\eta\vartheta} & 0 \\ L_{\vartheta\eta} & L_{\vartheta\vartheta} & 0 \\ L_{\tau\eta} & L_{\tau\vartheta} & \Psi \end{vmatrix} . \tag{3.6.33}$$

From the above equations we get

$$\begin{aligned} \varphi_r &= \frac{\partial\eta}{\partial r} = -\frac{\partial^2}{\partial r\partial z}(\Omega\Psi) , \\ u_\vartheta &= \frac{\partial\vartheta}{\partial r} = 2\alpha\frac{\partial}{\partial r}(\Xi\Psi) , \\ \varphi_z &= \tau = \Phi\psi - \frac{\partial^2}{\partial z^2}(\Omega\Psi) . \end{aligned} \tag{3.6.34}$$

where

$$\Omega = (\mu+\alpha)(\beta+\gamma-\varepsilon)\nabla^2 - 4\alpha^2 , \quad \Phi = (\mu+\alpha)\nabla^2[(\beta+2\gamma)\nabla^2 - 4\alpha^2] ,$$

$$\Xi = (\beta+2\gamma)\nabla^2 - 4\alpha .$$

By substituting (3.6.34) into the third relation of Eq.(3.6.32), we obtain the equation

$$[(\beta+2\gamma)\nabla^2-4\alpha]\nabla^2[(\mu+\alpha)(\gamma+\varepsilon)\nabla^2-4\alpha\mu]\Psi(r,z)+Y_z(r,z)=0. \tag{3.6.35}$$

Eq.(3.6.35) is an analogon with Eq.(3.6.18). There is no counterpart of this equation in the classical theory of elasticity. Putting $\alpha=0$ in the relations (3.6.34) and (3.6.35) we obtain the equation

$$(\beta+2\gamma)(\gamma+\varepsilon)\nabla^2\nabla^2\varphi+Y_z=0 \quad , \quad \varphi=\mu\nabla^2\Psi \quad , \tag{3.6.36}$$

and the relations

$$\begin{aligned} \varphi_r &= -(\beta+\gamma-\varepsilon)\frac{\partial^2\varphi}{\partial r\partial z} \\ \mu_\theta &= 0 \; , \\ \varphi_z &= (\beta+2\gamma)\nabla^2\varphi-(\beta+\gamma-\varepsilon)\frac{\partial^2\varphi}{\partial z^2} \; . \end{aligned} \tag{3.6.37}$$

Eq.(3.6.36) and the relations (3.6.37) refer to a quasi-elastic body in which the rotation $\underline{\varphi}$ is the only one possible. The state of displacement $\underline{u}\equiv(0,u_\theta,0)$ and rotations $\underline{\varphi}\equiv(\varphi_r,0,\varphi_z)$ is connected with the state of stress

$$\underline{\sigma}=\begin{vmatrix} 0 & \sigma_{r\theta} & 0 \\ \sigma_{\theta r} & 0 & \sigma_{\theta z} \\ 0 & \sigma_{z\theta} & 0 \end{vmatrix} \quad , \qquad \underline{\mu}=\begin{vmatrix} \mu_{rr} & 0 & \mu_{rz} \\ 0 & \mu_{\theta\theta} & 0 \\ \mu_{zr} & 0 & \mu_{zz} \end{vmatrix} \quad , \tag{3.6.38}$$

where

$$
\begin{aligned}
\sigma_{r\theta} &= \mu\left(\frac{\partial u_\theta}{\partial r} - \frac{u_\theta}{r}\right) + \frac{\alpha}{r}\frac{\partial}{\partial r}(r u_\theta) - 2\alpha\varphi_z , \\
\sigma_{\theta r} &= \mu\left(\frac{\partial u_\theta}{\partial r} - \frac{u_\theta}{r}\right) - \frac{\alpha}{r}\frac{\partial}{\partial r}(r u_\theta) + 2\alpha\varphi_z , \\
\sigma_{\theta z} &= \mu\frac{\partial u_\theta}{\partial z} - \frac{\alpha}{r}\frac{\partial}{\partial z}(r u_\theta) - 2\alpha\varphi_r , \\
\sigma_{z\theta} &= \mu\frac{\partial u_\theta}{\partial z} + \frac{\alpha}{r}\frac{\partial}{\partial z}(r u_\theta) + 2\alpha\varphi_r ,
\end{aligned}
\tag{3.6.39}
$$

and

$$
\begin{aligned}
&\mu_{rr} = 2\gamma\frac{\partial\varphi_r}{\partial r} + \beta\varkappa , \quad \mu_{\theta\theta} = 2\gamma\frac{\varphi_r}{r} + \beta\varkappa , \\
&\mu_{zz} = 2\gamma\frac{\partial\varphi_z}{\partial z} + \beta\varkappa , \\
&\mu_{rz} = \gamma\left(\frac{\partial\varphi_z}{\partial r} + \frac{\partial\varphi_r}{\partial z}\right) - \varepsilon\left(\frac{\partial\varphi_r}{\partial z} - \frac{\partial\varphi_z}{\partial r}\right) \\
&\mu_{zr} = \gamma\left(\frac{\partial\varphi_z}{\partial r} + \frac{\partial\varphi_r}{\partial z}\right) + \varepsilon\left(\frac{\partial\varphi_r}{\partial z} - \frac{\partial\varphi_z}{\partial r}\right) .
\end{aligned}
\tag{3.6.40}
$$

By expressing the above stresses in function Ψ , we have

$$
\begin{aligned}
\sigma_{r\theta} &= 2\alpha\left[(\mu+\alpha)\frac{\partial^2}{\partial r^2} - (\mu-\alpha)\frac{1}{r}\frac{\partial}{\partial r}\right](\Xi\Psi) - 2\alpha\left[\Phi\Psi - \frac{\partial^2}{\partial z^2}(\Omega\Psi)\right] , \\
\sigma_{\theta r} &= 2\alpha\left[(\mu-\alpha)\frac{\partial^2}{\partial r^2} - (\mu+\alpha)\frac{1}{r}\frac{\partial}{\partial r}\right](\Xi\Psi) - 2\alpha\left[\Phi\Psi - \frac{\partial^2}{\partial z^2}(\Omega\Psi)\right] , \\
\sigma_{z\theta} &= 2\alpha\frac{\partial^2}{\partial r\partial z}\left[(\mu+\alpha)(\Xi\Psi) - \Omega\Psi\right] , \\
\sigma_{\theta z} &= 2\alpha\frac{\partial^2}{\partial r\partial z}\left[(\mu-\alpha)(\Xi\Psi) + \Omega\Psi\right] ,
\end{aligned}
\tag{3.6.41}
$$

and

$$\mu_{rr} = -\frac{\partial}{\partial z}\left[2\gamma\frac{\partial^2}{\partial r^2}(\Omega\Psi) - \beta(\Phi - \nabla^2\Omega)\Psi\right],$$

$$\mu_{\theta\theta} = -\frac{\partial}{\partial z}\left[2\gamma\frac{1}{r}\frac{\partial}{\partial r}(\Omega\Psi) - \beta(\Phi - \nabla^2\Omega)\Psi\right],$$

$$\mu_{zz} = -\frac{\partial}{\partial z}\left[2\gamma\left(\frac{\partial^2\Omega}{\partial z^2} - \Phi\right)\Psi - \beta(\Phi - \nabla^2\Omega)\Psi\right], \tag{3.6.42}$$

$$\mu_{zr} = (\gamma - \varepsilon)\frac{\partial}{\partial r}\left(\Phi - \frac{\partial^2\Omega}{\partial z^2}\right)\Psi - (\gamma + \varepsilon)\frac{\partial^3}{\partial r\partial z^2}(\Omega\Psi),$$

$$\mu_{rz} = (\gamma + \varepsilon)\frac{\partial}{\partial r}\left(\Phi - \frac{\partial^2\Omega}{\partial z^2}\right)\Psi - (\gamma - \varepsilon)\frac{\partial^3}{\partial r\partial z^2}(\Omega\Psi).$$

Let us consider an example of the application of the function Ψ. Suppose a halfspace $z \geq 0$, loaded in the plane $z = 0$ by the twisting moments $m(r)$ with vectors directed parallel to the z-axis.

The function Ψ will be expressed in the form of Hankel's integral

$$\Psi(r,\xi) = \int_0^\infty Z(z) J_0(r\xi)\, d\xi, \tag{3.6.43}$$

where

$$Z(z) = Ae^{-\xi z} + Be^{-\delta z} + Ce^{-\nu z}.$$

Here

$$\delta = (\xi^2 + a^2)^{1/2}, \qquad \nu = (\xi^2 + \sigma^2)^{1/2},$$

and

$$a^2 = \frac{4\alpha}{\beta + 2\gamma} \quad , \qquad \sigma^2 = \frac{4\mu\alpha}{(\gamma+\varepsilon)(\mu+\alpha)}$$

The quantities A , B , C , functions of parameter ξ , will be determined from the boundary conditions

$$\mu_{zz}(r,0) = -m(r) \ , \ \mu_{zr}(r,0) = 0 \ , \ \sigma_{z\theta}(r,0) = 0. \quad (3.6.44)$$

The stresses μ_{zz}, μ_{zr}, $\sigma_{z\theta}$ are expressed in function Ψ by the formulae (3.6.41) and (3.6.42).

Chapter 4.

Problems of Thermoelasticity.

4. 1 Fundamental Relations and Equations.

The dynamical problem of thermoelasticity +) for an elastic, isotropic, and centrosymmetric body consists in determining the stresses $\sigma_{ji}(\underline{x},t)$, $\mu_{ji}(\underline{x},t)$ and deformations $\gamma_{ji}(\underline{x},t)$ $\varkappa_{ji}(\underline{x},t)$ of the classes $C^{(1)}$ and the functions $\underline{\dot{u}}(\underline{x},t)$, $\varphi(\underline{x},t)$ and temperature $\vartheta(\underline{x},t)$ of the class $C^{(2)}$, for $\underline{x} \in V + A$.

These functions satisfy:

a) the equilibrium equations

(4.1.1) $$\sigma_{ji,j} = \rho \ddot{u}_i - X_i(\underline{x},t) \quad ,$$

(4.1.2) $$\epsilon_{ijk}\sigma_{jk} + \mu_{ji,j} = J\ddot{\varphi}_i(\underline{x},t) - Y_i \;, \quad \underline{x} \in V \,, \; t>0$$

b) the linearized relations between the state of stresses and the state of deformations

(4.1.3) $$\sigma_{ji} = (\mu+\alpha)\gamma_{ji} + (\mu-\alpha)\gamma_{ij} + (\lambda\gamma_{kk} - \nu\vartheta)\delta_{ji} \;,$$

(4.1.4) $$\mu_{ji} = (\gamma+\varepsilon)\varkappa_{ji} + (\gamma-\varepsilon)\varkappa_{ij} + \beta\varkappa_{kk}\delta_{ji} \;,$$

+) W.Nowacki, "Couple-stresses in the theory of thermoelasticity", Proc. of IUTAM-Symposium on Irreversible Aspects in Continuum Mechanics, Vienna, 1968, Springer Verlag, Wien - - New York, 1968.

where

$$\gamma_{ji} = u_{i,j} - \epsilon_{kji}\varphi_k , \qquad (4.1.5)$$

$$\varkappa_{ji} = \varphi_{i,j} . \qquad (4.1.6)$$

c) the equation of thermal conductivity, and

d) the boundary conditions

$$\sigma_{ji}n_j = p_i(\underline{x}, t), \ \mu_{ji}n_j = m_i(\underline{x}, t), \ \theta = \vartheta(\underline{x}, t), \ \underline{x}\in A_\sigma \qquad (4.1.7)$$

$$u_i = f_i(\underline{x}, t), \ \varphi_i = g_i(\underline{x}, t), \ \theta = \vartheta(\underline{x}, t), \underline{x}\in A_u, t>0, \qquad (4.1.8)$$

e) the initial conditions

$$u_i = k_i(\underline{x}) \quad , \quad \dot{u}_i = h_i(\underline{x}) \quad , \quad \theta = s(\underline{x}) \qquad (4.1.9)$$

$$\varphi_i = l_i(x) \quad , \quad \dot{\varphi}_i = s_i(\underline{x}) \quad , \quad \underline{x}\in V, \ t=0. \qquad (4.1.10)$$

We did not yet present the equation of thermal conductivity. This equation can be derived by taking into account the relations obtained by formulas (1.3.17)

$$T\dot{S} = k\nabla^2\theta \ , \qquad T = T_0 + \theta \ , \qquad (4.1.11)$$

and Eq.(1.4.11)

$$S = \nu\gamma_{kk} + c_\varepsilon \log\frac{T}{T_0} \qquad , \qquad \nu = (3\lambda + 2\mu)\alpha_t. \qquad (4.1.12)$$

k , in Eq.(1.1.11), denotes a coefficient of thermal conductivity. By eliminating $\dot{S}$ between Eqs.(4.1.11) and (4.1.12), we obtain

$$\nabla^2\theta - \frac{1}{\varkappa}\dot{\theta} - \eta\left(1 + \frac{\theta}{T_0}\right)\operatorname{div}\underline{\dot{u}} = 0 , \ \varkappa = \frac{k}{c_\varepsilon} , \ \eta = \frac{\nu T_0}{k} . \qquad (4.1.13)$$

This is a nonlinear equation and can be linearized by the assumption, which we adopt, that $\left|\frac{\vartheta}{T_0}\right| \ll 1$.

By taking into account that heat sources act in the body and denoting by W the heat referred to the unit of volume and time, we obtain the following generalized equation of heat conduction

$$(4.1.14) \qquad \nabla^2\vartheta - \frac{1}{\varkappa}\dot{\vartheta} - \eta \operatorname{div} \dot{\underline{u}} = -\frac{Q}{\varkappa} \;, \qquad Q = \frac{W}{k} \,.$$

Observe that only dilatation term $\gamma_{kk} = \operatorname{div} \underline{u}$ occurs in this equation while the expression $\varkappa_{kk} = \operatorname{div} \underline{\varphi}$ does not. Eq.(4.1.14) is of exactly the same form as in the case of the symmetric elasticity.

Let us pass to the representation of the differential equations of the problem choosing as the unknow functions the displacements $\underline{u}(\underline{x},t)$, the rotations $\underline{\varphi}(\underline{x},t)$ and the temperature $\vartheta(\underline{x},t)$. By eliminating the stresses σ_{ji} and μ_{ji} between Eqs.(4.1.1) and (4.1.2), by means of relations (1.1.3) and (1.1.4), expressing them by the functions $\underline{u}$ and $\underline{\varphi}$ (formulae (4.1.5) and (4.1.6)) we obtain together with Eq.(4.1.13) the following system of differential equations

$$(4.1.15) \qquad \Box_2 \underline{u} + (\lambda + \mu - \alpha)\operatorname{grad}\operatorname{div} \underline{u} + 2\alpha \operatorname{rot} \underline{\varphi} + \underline{X} = \nu \operatorname{grad} \vartheta \,,$$

$$(4.1.16) \qquad \Box_4 \underline{\varphi} + (\beta + \gamma - \varepsilon)\operatorname{grad}\operatorname{div} \underline{\varphi} + 2\alpha \operatorname{rot} \underline{u} + \underline{Y} = 0 \,,$$

$$(4.1.17) \qquad \nabla^2\vartheta - \frac{1}{\varkappa}\dot{\vartheta} - \eta \operatorname{div} \dot{\underline{u}} = -\frac{Q}{\varkappa} \,.$$

We have obtained a coupled system of seven equations of seven unknowns, namely, three components of the displacement $\underline{u}$, three components of the rotation $\underline{\varphi}$ and the temperature ϑ. These fields can be generated by loadings, a heating of a surface, body forces and moments, and heat sources. The boundary and initial conditions, connected with Eqs.(4.1.15) ÷ (4.1.17), were given above (formulae (4.1.7) ÷ (4.1.10)).

If we assume in Eqs.(4.1.15) and (4.1.16) that $\alpha = 0$, then we obtain the following system of equations

$$\mu\nabla^2\underline{u} + (\lambda+\mu)\,\mathrm{grad\,div}\,\underline{u} + \underline{X} = \rho\ddot{\underline{u}} + \nu\,\mathrm{grad}\,\vartheta, \qquad (4.1.18)$$

$$\nabla^2\vartheta - \frac{1}{\varkappa}\dot{\vartheta} - \eta\,\mathrm{div}\,\dot{\underline{u}} = -\frac{Q}{\varkappa}, \qquad (4.1.19)$$

and

$$(\gamma+\varepsilon)\nabla^2\underline{\varphi} + (\gamma+\beta-\varepsilon)\,\mathrm{grad\,div}\,\underline{\varphi} + \underline{Y} = \mathfrak{J}\ddot{\underline{\varphi}}. \qquad (4.1.20)$$

We recognize in Eqs.(4.1.18) and (4.1.19) the equations of the classical thermoelasticity+). Eq.(4.1.20) refers to a hypothetical medium with only rotations possible. It is interesting to note that the temperature ϑ joins first two equations only of the system. The properties of both the media in Eqs.(4.1.15) ÷ (4.1.18) are coupled, each particle is subject to the displacement $\underline{u}$, the rotation $\underline{\varphi}$, and the increment of tem

+) W. Nowacki, "Dynamical problems of thermoelasticity" (in Polish), PWN, Warsaw, 1968.

perature ϑ. Here the quantities $\underline{u}$, $\underline{\varphi}$, and ϑ are independent of one another.

Let us perform on Eqs.(4.1.15) and (4.1.16) the operation of divergence and add to the thus-obtained equations the equation of thermal conductivity (4.1.17). We obtain the equations

$$(4.1.21)\qquad \begin{aligned} &\left(\nabla^2 - \frac{1}{c_1^2}\partial_t^2\right)\gamma_{kk} + \frac{1}{c_1^2\rho}\operatorname{div}\underline{X} = m\nabla^2\vartheta,\\ &\left(\nabla^2 - \frac{1}{\varkappa}\partial_t\right)\vartheta - \eta\dot{\gamma}_{kk} = -\frac{Q}{\varkappa}, \end{aligned}$$

and

$$(4.1.22)\qquad \left(\nabla^2 - \frac{1}{c_3^2}\partial_t^2 - \frac{\nu_0^2}{c_3^2}\right)\varkappa_{kk} + \frac{1}{Jc_3^2}\operatorname{div}\underline{Y} = 0 .$$

Here

$$\gamma_{kk} = \operatorname{div}\underline{u}, \quad \varkappa_{kk} = \operatorname{div}\underline{\varphi}, \quad m = \frac{\nu}{\lambda + 2\mu}, \quad \nu_0^2 = \frac{4\alpha}{J} .$$

Here it is apparent that temperature is connected with the dilatation.

By eliminating temperature from Eqs.(4.1.21), we obtain the following form of wave equation for the dilatation

$$(4.1.23)\qquad \begin{aligned} &\left[\left(\nabla^2 - \frac{1}{\varkappa}\partial_t\right)\left(\nabla^2 - \frac{1}{c_1^2}\partial_t^2\right) - m\eta_0\partial_t\nabla^2\right]\gamma_{kk} =\\ &\qquad = -\frac{mQ}{\varkappa} - \frac{1}{c_1^2\rho}\left(\nabla^2 - \frac{1}{\varkappa}\partial_t\right)\operatorname{div}\underline{X} . \end{aligned}$$

Eq.(4.1.23) does not differ in our form from the analogous one for the dilatational wave in the theory of symmetric elasticity. Let us note that in an infinite elastic space temperature can produce only a dilatational wave, here Eq.(4.1.23) is independent of temperature.

If we perform on Eqs.(4.1.15) and (4.1.16) the operation of rot, then we obtain the following equations

$$\square_2 \underline{S} + 2\alpha \operatorname{rot} \underline{T} + \operatorname{rot} \underline{X} = 0 , \qquad (4.1.24)$$

$$\square_4 \underline{T} + 2\alpha \operatorname{rot} \underline{S} + \operatorname{rot} \underline{Y} = 0 , \qquad (4.1.25)$$

where

$$\underline{S} = \operatorname{rot} \underline{u} , \quad \underline{T} = \operatorname{rot} \underline{\varphi} .$$

These equations are identical with Eqs.(1.7.15) and (1.7.16). Temperature does not appear in these equations. Therefore, in an infinite space the waves can be produced by the quantities rot $\underline{X}$ and rot $\underline{Y}$; however, not by heat sources or initial conditions for temperature.

Let us consider a particular case in which the causes producing the motion of body change in time very slowly. Then the inertial terms in the equations of motion can be neglected and the problem can be regarded as a quasi-statical one. We obtain the couple system of equations (for $\underline{X} = 0$, $\underline{Y} = 0$)

$$(\mu + \alpha)\nabla^2 \underline{u} + (\lambda + \mu - \alpha)\operatorname{grad}\operatorname{div} \underline{u} + 2\alpha \operatorname{rot} \underline{\varphi} = \nu \operatorname{grad} \theta, \qquad (1.1.26)$$

$$(\gamma + \varepsilon)\nabla^2 \underline{\varphi} - 4\alpha \underline{\varphi} + (\beta + \gamma - \varepsilon)\operatorname{grad}\operatorname{div} \underline{\varphi} + 2\alpha \operatorname{rot} \underline{u} = 0, \qquad (4.1.27)$$

$$\nabla^2 \theta - \frac{1}{\varkappa}\dot{\theta} - \eta \operatorname{div} \dot{\underline{u}} = -\frac{Q}{\varkappa} . \qquad (4.1.28)$$

If we must deal with a stationary heat flow, then, the time derivatives in Eq.(4.1.28) disappear and all the quantities are independent of time. The remaining system of equations takes the following form

(4.1.29) $$(\mu+\alpha)\nabla^2\underline{u} + (\lambda+\mu-\alpha)\,\mathrm{grad\,div}\,\underline{u} + 2\alpha\,\mathrm{rot}\,\underline{\varphi} = \nu\,\mathrm{grad}\,\vartheta ,$$

(4.1.30) $$(\gamma+\varepsilon)\nabla^2\underline{\varphi} - 4\alpha\underline{\varphi} + (\beta+\gamma-\varepsilon)\,\mathrm{grad\,div}\,\underline{\varphi} + 2\alpha\,\mathrm{rot}\,\underline{u} = 0 ,$$

(4.1.31) $$\nabla^2\vartheta = -\frac{Q}{\varkappa} .$$

The equation of thermal conductivity has become the equation of Poisson type. We determine the temperature ϑ from Eq.(4.1.31) and substitute it, as a known function, in Eq.(4.1.29). Only Eqs. (4.1.29) and (4.1.30) are coupled. The easiest way to get the solution to the system of equations is to introduce the potential of thermoelastic displacement

(4.1.32) $$\underline{u}' = \mathrm{grad}\,\Phi$$

and to assume that $\underline{\varphi}' = 0$.

By substituting (4.1.32) in Eqs.(4.1.29) and (4.1.30), we reduce this system of equations to the Poisson equation

(4.1.33) $$\nabla^2\Phi = m\vartheta , \qquad m = \frac{\nu}{\lambda+2\mu} , \qquad \underline{\varphi}' = 0 .$$

We express the stresses by means of the function Φ

(4.1.34) $$\sigma'_{ij} = \sigma'_{ji} = 2\mu(\Phi_{,ij} - \delta_{ij}\nabla^2\Phi) , \qquad \mu'_{ij} = 0 .$$

These quantities constitute the complete solution in an infinite space and are identical for Hooke's medium and for Cosserat's medium. If the region is bounded, we add to the stresses σ'_{ij}, μ'_{ij} such stresses σ''_{ij}, μ''_{ij} that any boundary conditions on the boundary are satisfied. The stresses σ''_{ij}, μ''_{ij} are connected with such state of displacements and rotations $\underline{u}''$, $\underline{\varphi}''$ that satisfies the following homogeneous system of equations

$$(\mu+\alpha)\nabla^2\underline{u}''+(\lambda+\mu-\alpha)\,\mathrm{grad\,div}\,\underline{u}''+2\alpha\,\mathrm{rot}\,\underline{\varphi}''=0 \,, \qquad (4.1.35)$$

$$(\gamma+\varepsilon)\nabla^2\underline{\varphi}''-4\alpha\underline{\varphi}''+(\beta+\gamma-\varepsilon)\,\mathrm{grad\,div}\,\underline{\varphi}''+2\alpha\,\mathrm{rot}\,\underline{u}''=0 \,. \qquad (4.1.36)$$

Let us call our attention to two very simple states of deformation. Assume that in an infinite simple-connected body a constant temperature acts on a rigidly fixed surface A ($\underline{u}=0$, $\underline{\varphi}=0$). Then Eq.(4.1.29) becomes a homogeneous one, and consequently for the assumed homogeneous boundary conditions we obtain $\underline{u}=0$, $\underline{\varphi}=0$ at any point of the body. Thus $\gamma_{ij}=0$ and $\varkappa_{ij}=0$. It results from relations (4.1.3) and (4.1.4) that

$$\sigma_{ij}=\sigma_{(ij)}=-\nu\vartheta\delta_{ij} \,, \qquad \mu_{ij}=0 \,. \qquad (4.1.37)$$

Next let us consider a simple-connected body in which there exist a field of temperature $\vartheta(\underline{x})$. Now the following question arises: what kind of temperature distribution can exist in the body that neither stresses σ_{ji} nor moment stresses μ_{ji} appear. We obtain from relations (1.4.5)

$$\gamma_{ji}=\alpha_t\vartheta\delta_{ji} \,, \qquad \varkappa_{ji}=0 \,.$$

It still remains to satisfy the compatibility equations (formulae (1.5.9))

$$(4.1.38)\qquad \begin{aligned} &\gamma_{li,h} - \gamma_{hi,l} + \epsilon_{kih}\varkappa_{lk} - \epsilon_{kil}\varkappa_{hk} = 0 \\ &\varkappa_{lk,h} = \varkappa_{hk,l} . \end{aligned}$$

The second system of equations is satisfied immediately, the first one leads to the system of equations

$$\vartheta_{,ij} = 0 , \left(\text{no sum for } \quad i = j\right) .$$

Eqs.(4.1.38) hold when the temperature distribution is linear

$$(4.1.39)\qquad \vartheta = a_0 + x_i a_i \quad .$$

Here we obtained the same result as in the classical thermoelasticity.

Now, let us devote a few words to the so called "theory of thermal stresses". In this simplified theory we assume that the effect of the term $-\eta_0 \operatorname{div} \dot{\underline{u}}$ on the variation of the displacements $\underline{u}$, the rotations $\underline{\varphi}$, and the temperature ϑ is very small. Thus, the temperature can be determined from the simplified, classical equation of thermal conductivity

$$(4.1.40)\qquad \left(\nabla^2 - \frac{1}{\varkappa}\partial_t\right)\vartheta = -\frac{Q}{\varkappa} \quad .$$

Then, the temperature occurring in Eq.(1.1.15) can be regarded as a known function.

In this way we obtain a significant mathematical simplification.

Consequently, the temperature becomes an external source with respect to the body displacement.

Let us also consider an interesting relation referring to a change of the body volume produced by a stationary field of temperature. Multiply the equilibrium equations

$$\sigma_{ji,j} + X_i = 0 \quad , \quad \epsilon_{ijk}\sigma_{jk} + \mu_{ji,j} + Y_i = 0 \ ,$$

by V and integrate over the region x_i of a simply-connected body:

$$\int_V (\sigma_{ji,j} + X_i) x_i \, dV = 0$$

$$\int_V (\epsilon_{ijk}\sigma_{jk} + \mu_{ji,j} + Y_i) x_i \, dV = 0 .$$

By taking into account that $p_i = \sigma_{ji} n_j$, $m_i = \mu_{ji} n_j$ and applying the divergence theorem, we obtain

$$\begin{aligned} &\int_A p_i x_i \, dA + \int_V X_i x_i \, dV = \int_V \sigma_{kk} \, dV \ , \\ &\int_A m_i x_i \, dA + \int_V Y_i x_i \, dV + \int_V \epsilon_{ijk}\sigma_{jk} x_i \, dV = \int_V \mu_{kk} \, dV . \end{aligned} \qquad (4.1.41)$$

Since

$$\sigma_{kk} = 3K\gamma_{kk} - 3\nu\vartheta \quad , \quad \mu_{kk} = 3L\varkappa_{kk} \quad ,$$

$$K = \lambda + \frac{2}{3}\mu \quad , \quad L = \beta + \frac{2}{3}\gamma \quad ,$$

then

$$\begin{aligned} &\int_V \gamma_{kk} \, dV = \frac{1}{3K}\left(\int_A p_i x_i \, dA + \int_V X_i x_i \, dV\right) + 3\alpha_t \int_V \vartheta \, dV \ , \\ &\int_V \varkappa_{kk} \, dV = \frac{1}{3L}\left(\int_A m_i x_i \, dA + \int_V Y_i x_i \, dV + \int_V \epsilon_{ijk} x_i \sigma_{jk} \, dV\right) . \end{aligned} \qquad (4.1.42)$$

The integral $\int_V \gamma_{kk} dV = \Delta V$ represents the increment of the body volume. We observe from the first Eq.(4.1.42) that this increment depends only on loadings p_i, body forces X_i and an increment of temperature ϑ. The second formula represents the integral of the quantity $\varkappa_{kk} = \operatorname{div} \underline{\varphi}$ taken over the body volume. Since $\int_V \varkappa_{kk} dV = \int_A \varphi_j n_j dA$, then, the integral can be interpreted as the integral of the normal component of the vector $\underline{\varphi}$ over the surface A of the body V. Since

$$\epsilon_{ijk} x_i \sigma_{jk} = \epsilon_{ijk} x_i \sigma_{<jk>}$$

and $\sigma_{<jk>} = 2\alpha \gamma_{<jk>}$, then,

$$(4.1.43) \qquad \int_V \varkappa_{kk} dV = \frac{1}{3L}\left(\int_A m_i x_i dA + \int_V Y_i x_i dV + 2\alpha \int_V \epsilon_{ijk} x_i \gamma_{<jk>} dV \right).$$

This integral depends on mass moments, moments distributed over the surface A and the antisymmetric part of the strain tensor. If the temperature only acts on a simply-connected body, then,

$$(4.1.44) \qquad \Delta V = 3\alpha_t \int_V \vartheta dV .$$

On the other hand, we obtain from the first Eq.(4.1.41) for $X_i = 0$, $Y_i = 0$

$$(4.1.45) \qquad \int_V \sigma_{kk} dV = 0 .$$

Formulae (4.1.44) and (4.1.45) are identical with those obtained within the frames of the classical theory of elasticity.

The formulae, here presented, can be derived in a

more general way from the reciprocity theorem (See Sect.4.5).

4. 2 Potentials and Stress Functions of Thermoelasticity.

The dynamic equations of thermoelasticity

$$\Box_2 \underline{u} + (\lambda + \mu - \alpha)\,\mathrm{grad}\,\mathrm{div}\,\underline{u} + 2\alpha\,\mathrm{rot}\,\underline{\varphi} + \underline{X} = \nu\,\mathrm{grad}\,\vartheta,$$
$$\Box_4 \underline{\varphi} + (\beta + \gamma - \varepsilon)\,\mathrm{grad}\,\mathrm{div}\,\underline{\varphi} + 2\alpha\,\mathrm{rot}\,\underline{u} + \underline{Y} = 0 \qquad (4.2.1)$$

can be separated in two different ways. First way, analogous to Lamé's procedure applied in classic elastokinetics, consists in the decomposition of the vectors $\underline{u}$ and $\underline{\varphi}$ in potential and solenoidal parts, respectively

$$\underline{u} = \mathrm{grad}\,\Phi + \mathrm{rot}\,\underline{\Psi}, \qquad \mathrm{div}\,\underline{\Psi} = 0,$$
$$\underline{\varphi} = \mathrm{grad}\,\Sigma + \mathrm{rot}\,\underline{H}, \qquad \mathrm{div}\,\underline{H} = 0. \qquad (4.2.2)$$

By decomposing in a similar way the expressions for the body forces and the body-couples

$$\underline{X} = \rho(\mathrm{grad}\,\vartheta + \mathrm{rot}\,\underline{\chi}), \qquad \mathrm{div}\,\underline{\chi} = 0,$$
$$\underline{Y} = \mathfrak{J}(\mathrm{grad}\,\sigma + \mathrm{rot}\,\underline{\eta}), \qquad \mathrm{div}\,\underline{\eta} = 0, \qquad (4.2.3)$$

and substituting (4.2.2) and (4.2.3) to the equations of motion (4.2.1), we obtain the following system of equations

$$\Box_1 \Phi + \rho\vartheta = \nu\theta, \qquad (4.2.4)$$

$$\Box_3 \Sigma + \mathfrak{J}\sigma = 0, \qquad (4.2.5)$$

(4.2.6) $$(\square_2\square_4 + 4\alpha^2\nabla^2)\underline{\Psi} = 2\alpha\mathfrak{I}\,\mathrm{rot}\,\underline{\eta} - \varrho\square_4\underline{\chi}\,,$$

(4.2.7) $$(\square_2\square_4 + 4\alpha^2\nabla^2)\underline{H} = 2\alpha\varrho\,\mathrm{rot}\,\underline{\chi} - \mathfrak{I}\square_2\underline{\eta}\,.$$

To these equations it is necessary to add a heat conduction equation

(4.2.8) $$D\vartheta - \eta\partial_t\nabla^2\Phi = -\frac{Q}{\varkappa}\,, \qquad D = \nabla^2 - \frac{1}{\varkappa}\partial_t\,.$$

Let us notice, that Eqs.(4.2.4), (4.2.8) and (4.2.6) (4.2.7) are mutually coupled.

By eliminating from Eqs.(4.2.4) and (4.2.8), first a temperature, and second the function Φ , we have

(4.2.9) $$M\Phi = -\frac{\nu}{\varkappa}Q - \varrho D\vartheta\,,$$

(4.2.10) $$M\vartheta = -\eta\varrho\,\partial_t\nabla^2\vartheta - \frac{1}{\varkappa}\square_1 Q$$

where

$$M = \square_1 D - \nu\eta\partial_t\nabla^2\,.$$

We shall consider first the propagation of thermoelastic waves in an infinite space.

If the quantities σ , $\underline{\chi}$, $\underline{\eta}$ and initial conditions of the functions Σ , $\underline{\Psi}$, $\underline{H}$ are equal to zero, then in an unbounded elastic space only dilatational waves will propagate. Equation (4.2.9), describing these waves is identical with that obtained for the elastic classical medium (with no couple-

-stresses). These waves are attenuated and dispersed.

Since

$$u_i = \Phi_{,i} \,, \qquad \varphi_i = 0 \,, \qquad \gamma_{ji} = \Phi_{,ji} \,, \qquad \varkappa_{ji} = 0 \,,$$

we have

$$\sigma_{(ij)} = 2\mu(\Phi_{,ij} - \delta_{ij}\Phi_{,kk}) + \rho\delta_{ij}(\ddot{\Phi} - \vartheta) \,, \quad \sigma_{\langle ij\rangle} = 0 \,, \quad \mu_{ij} = 0.$$

If Q , ϑ, $\underline{\chi}$, $\underline{\eta}$ are equal to zero and the initial conditions of the functions ϑ, Φ, $\underline{\Psi}$, $\underline{H}$ are homogeneous, then, in an inifinite medium only microrotational waves, described by Eq. (1.2.5), propagate. We have, namely,

$$u_i = 0 \,, \quad \varphi_i = \Sigma_{,i} \,, \quad \varkappa_{ji} = \Sigma_{,ij} \,, \quad \gamma_{(ji)} = 0 \,, \quad \gamma_{\langle ji\rangle} = -\epsilon_{kij}\Sigma_{,k} \,.$$

Only couple-stresses will appear in the medium forming the symmetric tensor

$$\mu_{(ij)} = 2\gamma\Sigma_{,ij} + \beta\delta_{ij}\Sigma_{,kk} \,, \quad \mu_{\langle ij\rangle} = 0 \,, \quad \sigma_{ij} = 0 \,, \quad \mathrm{div}\,\underline{u} = 0.$$

The propagation of these waves is not accompanied by a temperature field.

Finally, in the case when the quantities Q , ϑ, σ are equal zero and the initial conditions of the functions Φ , Σ , ϑ are homogeneous. Only transverse waves propagate in an infinite space (described by the Eqs.(4.2.6) and (4.2.7)). In an infinite medium these waves are not accompanied by a temperature field. Since $\mathrm{div}\,\underline{u} = 0$, they do not induce any changes

in the volume of the body.
In a finite medium all three kinds of waves, discussed here, appear Eqs.(4.2.4) $\div$ (4.2.7) are coupled by means of the boundary conditions.

The second way of separation of the system of equations (4.2.1) $\div$ (4.2.8) is analogous to that one which was used by N. Sandru+) in micropolar elastokinetics. Let us eliminate from Eqs.(4.2.1), first the displacement $\underline{u}$, and second, the rotations $\underline{\varphi}$. As a result we shall obtain

(4.2.11) $$\Omega\underline{u} + \operatorname{grad}\operatorname{div}\Gamma\underline{u} + \Box_4(\underline{X} - \nu\operatorname{grad}\vartheta) - 2\alpha\operatorname{rot}\underline{Y} = 0,$$

(4.2.12) $$\Omega\underline{\varphi} + \operatorname{grad}\operatorname{div}\Theta\underline{\varphi} + \Box_2\underline{Y} - 2\alpha\operatorname{rot}\underline{X} = 0,$$

(4.2.13) $$D\vartheta - \eta\partial_t\operatorname{div}\underline{u} = -\frac{Q}{\varkappa}$$

where

$$\Omega = \Box_2\Box_4 - 4\alpha^2\nabla^2, \qquad \Gamma = (\lambda + \mu - \alpha)\Box_4 - 4\alpha^2,$$
$$\Theta = (\beta + \gamma - \varepsilon)\Box_2 - 4\alpha^2.$$

Let us notice the fact that in Eq.(4.2.12) a temperature does not appear. Let us eliminate from Eqs.(4.2.11), (4.2.12) first the temperature ϑ and second, the displacements $\underline{u}$. In this way we obtain a new system of equations

(4.2.14) $$D\Omega\underline{u} + \operatorname{grad}\operatorname{div}N\underline{u} + D\Box_4\underline{X} - 2\alpha\operatorname{rot}D\underline{Y} + \frac{\nu}{\varkappa}\operatorname{grad}(\Box_4 Q) = 0,$$

+) N. Sandru, op. cit. p. 53 .

$$\Omega \underline{\varphi} + \text{grad div} \,\Theta\, \underline{\varphi} + \Box_2 \underline{Y} - 2\alpha \,\text{rot}\, \underline{X} = 0 \,, \qquad (4.2.15)$$

$$\Box_4 M \vartheta + \eta \partial_t \,\text{div}\, \Box_4 \underline{X} + \frac{1}{\varkappa} \Box_1 \Box_4 Q = 0 \,, \qquad (4.2.16)$$

where

$$N = \Gamma D - \nu \eta \partial_t \Box_4 \quad .$$

We shall assume the following representation for displacements $\underline{u}$ and rotations $\underline{\varphi}$:

$$\underline{u} = \Box_4 M \underline{\xi} - \text{grad div} \, N \underline{\xi} \quad , \qquad (4.2.17)$$

$$\underline{\varphi} = \Box_2 \Box_3 \underline{\eta} - \text{grad div} \,\Theta\, \underline{\eta} \quad . \qquad (4.2.18)$$

By inserting the above formulae into Eqs.(4.2.14), (4.2.15), one can obtain the following differential equations:

$$D\Omega \Box_4 M \underline{\xi} + D \Box_4 \underline{X} - 2\alpha \,\text{rot}\, D \underline{Y} + \frac{\nu}{\varkappa} \text{grad} \Box_4 Q = 0 , \qquad (4.2.19)$$

$$\Omega \Box_2 \Box_3 \underline{\eta} + \Box_2 \underline{Y} - 2\alpha \,\text{rot}\, \underline{X} = 0 \quad . \qquad (4.2.20)$$

Here the relations were used

$$\Omega = \Box_2 \Box_4 + 4\alpha^2 \nabla^2 = \Box_1 \Box_4 - \Gamma \nabla^2 = \Box_2 \Box_3 - \Theta \nabla^2 ,$$

$$D\Omega + N \nabla^2 = \Box_4 M \quad . \qquad (4.2.20')$$

In Eqs.(4.2.19),(4.2.20) the differential operations on the body-force and body-couples figure what is termed as an inconvenience when equations are **being** solved.

Let us enrich representation (4.2.17),(4.2.18) by

adding a rotational term and a gradient of a certain scalar function ω.

(4.2.21) $\underline{u} = \Box_4 M\underline{F} - \mathrm{grad}\,\mathrm{div} N\underline{F} - 2\alpha\,\mathrm{rot}\,\Box_3\underline{G} + \nu\,\mathrm{grad}\,\omega$,

(4.2.22) $\underline{\varphi} = \Box_2\Box_3\underline{G} - \mathrm{grad}\,\mathrm{div}\,\Theta\underline{G} - 2\alpha\,\mathrm{rot}\,N\underline{F}$,

(4.2.23) $\vartheta = \eta\partial_t\,\mathrm{div}\,\Omega\underline{F} + \Box_1\omega$.

By inserting Eqs.(4.2.21)÷(4.2.23) into the system of equations (4.2.14)÷(4.2.16), one can obtain the following wave equations:

(4.2.24) $$\Omega M\underline{F} + \underline{X} = 0 ,$$

(4.2.25) $$\Omega\Box_3\underline{G} + \underline{Y} = 0 ,$$

(4.2.26) $$M\omega + \frac{Q}{\varkappa} = 0 .$$

We have obtained the system of equations in which the body forces, body couples and heat sources appear separately. Let us notice that in an infinite, elastic space the assumption $\underline{X} = 0$ with homogeneous initial conditions for $\underline{F}$ leads to a corollary that $\underline{F} = 0$ in a whole space. The same corollary concerns the function $\underline{G}$ with $\underline{Y} = 0$ and ω with $Q = 0$.

Equations (4.2.24)÷(4.2.26) are particularly useful in a case of the singular solutions in an infinite, micropolar space.

Let us consider the homogeneous equations of ther

moelasticity (4.2.24) (4.2.26):

$$M\Omega\underline{F} = 0 \quad , \quad \Box_3\Omega\underline{G} = 0 \quad , \quad M\omega = 0. \qquad (4.2.27)$$

A solution of these equations may be presented on a basis of T.Boggio theorem in a form

$$\underline{F} = \underline{F}' + \underline{F}'' , \quad \underline{G} = \underline{G}' + \underline{G}'' , \qquad (4.2.28)$$

where "primed" functions satisfy the equations:

$$\begin{aligned} M\underline{F}' = 0 \quad &, \quad \Omega\underline{F}'' = 0 \quad , \\ \Box_3\underline{G}' = 0 \quad &, \quad \Omega\underline{G}'' = 0 \quad . \end{aligned} \qquad (4.2.29)$$

By substituting Eq.(4.2.29) into Eqs.(4.2.21) ÷ (4.2.23), we obtain

$$\begin{aligned} \underline{u} &= \Box_4 M\underline{F}'' - \mathrm{grad}\,\mathrm{div}\, N(\underline{F}' + \underline{F}'') - 2\alpha\,\mathrm{rot}\,\Box_3\underline{G}'' + \nu\,\mathrm{grad}\,\omega , \\ \underline{\varphi} &= \Box_2\Box_3\underline{G}'' - \mathrm{grad}\,\mathrm{div}\,\Theta(\underline{G}' + \underline{G}'') - 2\alpha\,\mathrm{rot}\, M\underline{F}'' \\ \vartheta &= \eta\,\partial_t\,\mathrm{div}\,\Omega\underline{F}' + \Box_1\omega \quad . \end{aligned} \qquad (4.2.30)$$

Using formula (4.2.20') and equations

$$(\Box_4 M - N\nabla^2)\underline{F}'' = D\Omega\underline{F}'' = 0$$

$$(\Box_2\Box_3 - \nabla^2\Theta)\underline{G}'' = \Omega\underline{G}'' = 0 \quad , \qquad \mathrm{rot}\,\mathrm{rot}\,\underline{U} = \mathrm{grad}\,\mathrm{div}\,\underline{U} - \nabla^2\underline{U} \quad ,$$

we shall present the relations (4.2.30) in a form

$$\underline{u} = -\,\mathrm{grad}\left[\mathrm{div}(N\underline{F}') - \nu\omega\right] - \mathrm{rot}\left[\mathrm{rot}(N\underline{F}'') + 2\alpha\Box_3\underline{\Psi}\right], \qquad (4.2.31)$$

and

$$\underline{\varphi} = -\operatorname{grad}[\operatorname{div} \Theta \underline{G}'] - \operatorname{rot}[\operatorname{rot}(\Theta \underline{G}'') + 2\alpha M \underline{F}''] .$$

By comparison of the above relations with representation (4.2.2), we obtain the following relations between potentials $\underline{\Psi}$, $\underline{H}$, Φ, Σ and stress functions $\underline{F}$, $\underline{G}$, ω:

$$\begin{aligned} \Phi &= -\operatorname{div} N\underline{F}' - \nu\omega , & \underline{\Psi} &= -\operatorname{rot} N\underline{F}'' - 2\alpha \Box_3 \underline{G}'' , \\ \Sigma &= -\operatorname{div} \Theta \underline{G}' , & \underline{H} &= \operatorname{rot} \Theta \underline{G}'' - 2\alpha M \underline{F}'' . \end{aligned} \tag{4.2.32}$$

It remains to be checked whether the functions Φ, Σ, $\underline{\Psi}$, $\underline{H}$, expressed by the functions $\underline{F}$, $\underline{G}$, ω, satisfy the homogeneous wave equations (4.2.4)÷(4.2.5) and relation

$$\Omega \underline{\Psi} = 0 , \quad \Omega \underline{H} = 0 \tag{4.2.33}$$

which is obtained from homogeneous Eqs.(4.2.6) and (4.2.7).

Let us perform an operation $\Box_1$ on the first Eq. (4.2.32). Then,

$$\Box_1 \Phi = -\operatorname{div} \Box_1 N\underline{F}' + \nu \Box_1 \omega .$$

By taking into account Eq.(4.2.23), we obtain

$$\Box_1 \Phi = -\operatorname{div} \Box_1 N\underline{F}' + \nu(\vartheta - \eta \partial_t \operatorname{div} \Omega \underline{F}') ,$$

Since

$$(\Box_1 N + \eta \partial_t \Omega) F' = MF' = 0 ,$$

then, the homogeneous wave equation

$$\Box_1 \Phi = \nu \vartheta \qquad (4.2.34)$$

will be satisfied.

The remaining equations are also satisfied in virtue of Eq.(4.2.29). Hence, we have by turn

$$\begin{aligned} \Box_3 \Sigma &= -\mathrm{div}\,\Theta\,\Box_3 \underline{G}' = 0 \\ \Omega \underline{\Psi} &= -\mathrm{rot}(N\,\Omega \underline{F}'') - 2\alpha\,\Box_3 \Omega \underline{G}'' = 0\,, \qquad (4.2.35) \\ \Omega \underline{H} &= -\mathrm{rot}(\Theta\,\Omega \underline{G}'') - 2\alpha M\,\Omega \underline{F}'' = 0\,. \end{aligned}$$

We have considered here the coupled equations of thermoelasticity. A significant approximation we shall obtain within a frame of so named "theory of thermal stresses", in which appear a term $\eta \partial_t \mathrm{div}\, \underline{u}$, coupling a heat conduction equation with equations of motion. In this case temperature is determined from parabolic equation and is introduced as a known function into the first Eq.(4.2.1).

In this way a thermoelastic problem is reduced to the elastic one. Instead of the body forces $\underline{X}$ it is necessary to apply the reduced body forces $\underline{X}^* = \underline{X} - \nu\, \mathrm{grad}\, \vartheta$.

4. 3 Virtual Work Principle.

It may be easily shown that the following equation holds

$$\int_V [(X_i - \rho\ddot{u}_i)\delta u_i + (Y_i - \mathfrak{J}\ddot{\varphi}_i)\delta\varphi_i]dV + \int_A (p_i\delta u_i + m_i\delta\varphi_i)dA =$$

$$= \int_V (\sigma_{ji}\delta\gamma_{ji} + \mu_{ji}\delta\varkappa_{ji})dV\,. \tag{4.3.1}$$

The left-hand side of this equation represents the variation of work of external forces, while the right-hand side equals that of internal forces.

By introducing into the right-hand side of Eq. (4.3.1) the constitutive relations

$$\sigma_{ij} = 2\mu\gamma_{(ij)} + 2\alpha\gamma_{\langle ij\rangle} + (\lambda\gamma_{kk} - \nu\theta)\delta_{ji}\,,$$

$$\mu_{ij} = 2\gamma\varkappa_{(ij)} + 2\varepsilon\varkappa_{\langle ij\rangle} + \beta\varkappa_{kk}\delta_{ij}\,, \tag{4.3.2}$$

we reduce Eq.(4.3.1) to the form:

$$\int_V [(X_i - \rho\ddot{u}_i)\delta u_i + (Y_i - \mathfrak{J}\ddot{\varphi}_i)\delta\varphi_i]dV + \int_A (p_i\delta u_i + m_i\delta\varphi_i)dA =$$

$$= \delta W_\varepsilon - \nu\int_V \theta\,\delta\gamma_{kk}\,dV \tag{4.3.3}$$

where

$$\delta W_\varepsilon = \int_V (2\mu\gamma_{(ij)}\delta\gamma_{(ij)} + 2\alpha\gamma_{\langle ij\rangle}\delta\gamma_{\langle ij\rangle} + 2\gamma\varkappa_{(ij)}\delta\varkappa_{(ij)} +$$
$$+ 2\varepsilon\varkappa_{\langle ij\rangle}\delta\varkappa_{\langle ij\rangle} + \lambda\gamma_{kk}\delta\gamma_{nn} + \beta\varkappa_{kk}\delta\varkappa_{nn})dV\,.$$

We have to supplement Eq.(4.3.3) by a further equation, since only four causes, namely X_i, p_i, Y_i, m_i appear in this equation in explicit form. Thus, we adjoin to Eq.(4.3.3) the relation

$$-\nu \int_V \vartheta \delta \gamma_{kk} dV = \int_A \vartheta n_i \delta H_i dA + \frac{c_\varepsilon}{T_0} \int_V \vartheta \delta \vartheta dV + \frac{T_0}{k} \int_V \dot{H}_i \delta H_i dV . \tag{4.3.4}$$

derived from the heat equation by M.A. Biot+).

The vector $\underline{H}$ in (4.3.4) is connected with the vector of heat flow $\underline{q}$ and the entropy S by the following relations

$$\underline{q} = T_0 \dot{\underline{H}} \quad , \qquad S = -\operatorname{div} \underline{H} . \tag{4.3.5}$$

By introducing (4.3.4) into Eq.(4.3.3), we have

$$\delta(\mathcal{W}_\varepsilon + \mathcal{P} + \mathcal{D}) = \int_V [(X_i - \varrho \ddot{u}_i)\delta u_i + (Y_i - \mathcal{J}\ddot{\varphi}_i)\delta \varphi_i] dV + \int_A (p_i \delta u_i + m_i \delta \varphi_i) dA - \int_A \vartheta n_i \delta H_i dA . \tag{4.3.6}$$

Here the heat potential $\mathcal{P}$ and the dissipation function $\mathcal{D}$ were applied, where

$$\mathcal{P} = \frac{c_\varepsilon}{2T_0} \int_V \vartheta^2 dV \quad , \qquad \delta \mathcal{D} = \frac{T_0}{k} \int_V \dot{H}_i \delta H_i dV .$$

for $\quad Y_i = 0 , \; m_i = 0 , \; \varkappa_{ji} = 0 , \; \alpha = \gamma = \varepsilon = \beta = 0$

reduces to the variational equation of the thermoelastic medium without couple-stresses.

The variational principle Eq.(4.3.6) may be used

+) Biot, M.A.: Thermoelasticity and irreversible thermodynamics. J.Appl. Phys. 27, 1965.

for the derivation of the energy theorem, if we compare the functions u_i, φ_i, ϑ at point $\mathbf{x}$ and time t with those actually occuring in the same point after a time interval dt. Thus, introducing into Eq.(4.3.6)

$$\delta u_i = v_i dt \;,\quad \delta\varphi_i = w_i dt \;,\quad \delta\vartheta = \dot{\vartheta} dt \;,\quad \delta H_i = \dot{H}_i dt = -\frac{k}{T_0}\vartheta_{,i} dt$$

and so on, we obtain the following formula

$$\frac{d}{dt}(\mathcal{K} + \mathcal{W}_\varepsilon + \mathcal{P}) + \chi_\vartheta = \int_V (X_i v_i + Y_i w_i) dV + \int_A (p_i v_i + m_i w_i) dA +$$

$$+ \frac{k}{T_0}\int_A \vartheta n_i \vartheta_{,i}\, dA \;, \tag{4.3.7}$$

where

$$\mathcal{K} = \frac{\varrho}{2}\int_V v_i v_i dV + \frac{\mathfrak{J}}{2}\int_V w_i w_i dV \;,\quad \chi_\vartheta = \frac{dD}{dt} = \frac{k}{T_0}\int_V \vartheta_{,i}\vartheta_{,i} dV \;.$$

Here $\mathcal{K}$ denotes kinetic energy, and χ_ϑ is proportional to the source of entropy which is always a positive quantity. The energy theorem (4.3.7) may be exploited to demonstrate the uniqueness theorem for a simply connected body. Such a demonstration may be carried out in a way similar to that indicated in Sect. 1.11.

Let the body be in the static equilibrium under the action of external forces and raising temperature.

The virtual work principle takes the form

$$\int_V (X_i \delta u_i + Y_i \delta\varphi_i) dV + \int_A (p_i \delta u_i + m_i \delta\varphi_i) dA =$$
$$= \delta \mathcal{W} - \nu \int_V \vartheta \delta \gamma_{kk} dV. \quad (4.3.8)$$

Since the body-forces and the body-couples as well as the tensions and moments of surface do not vary, we write Eq.(4.3.8) in the following form

$$\delta\Gamma = 0$$

where

$$\Gamma = \mathcal{W}_\varepsilon - \int_V (X_i u_i + Y_i \varphi_i) dV - \int_A (p_i u_i + m_i \varphi_i) dA +$$
$$- \nu \int_V \vartheta \gamma_{kk} dV. \quad (4.3.9)$$

Γ is the potential energy. Proceeding in an analogous way as for symmetric thermoelasticity, we arrive at the conclusion that Γ is minimum.

Let us go back to Eq.(4.3.8) and transform the last integral appearing in the right-hand part of this equation to the form

$$\nu \int_V \vartheta \delta \gamma_{kk} dV = \nu \int_V \vartheta \delta u_{k,k} dV =$$
$$= \nu \int_A \vartheta n_k \delta u_k dA - \nu \int_V \vartheta_{,k} \delta u_k dV. \quad (4.3.10)$$

After substituting (4.3.10) into (4.3.8) we obtain

$$\delta \mathcal{W}_\varepsilon = \int_V [(X_i - \gamma \vartheta_{,i}) \delta u_i + Y_i \delta\varphi_i] dV +$$
$$+ \int_A [(p_i + \nu \vartheta n_i) \delta u_i + m_i \delta \varphi_i] dA. \quad (4.3.11)$$

Now, we shall consider an identical body (i.e., of the same form and material), but be placed under isothermal conditions. Let the body-forces X_i^* and the body-couples Y_i^* act on the body. The tensions p_i^* and moments m_i^* are assumed to be given on the surface A_σ, while displacements u_i^* and rotations φ_i^* - on A_u. We ask the following question: What should be the quantities X_i^* and Y_i^* - expressing forces and couples acting inside the body - and, on the other hand, the quantities p_i^* and m_i^* - expressing the tensions and moments acting on the surface A_σ - with identical boundary conditions for A_u in order to obtain the same field of displacements u_i and rotations φ_i in both viz., thermoelastic and isothermal problems. To get the answer, we shall compare (4.3.11) with the virtual work equation

$$\delta W_\varepsilon = \int_V (X_i^* \delta u_i + Y_i^* \delta \varphi_i) dV + \int_A (p_i^* \delta u_i + m_i^* \delta \varphi_i) dA \; . \tag{4.3.12}$$

In view of the identity of u_i and φ_i fields, the left-hand parts of Eqs.(4.3.11) and (4.3.12) are identical, too. Thus, we obtain the following relations

$$\begin{aligned} X_i^* &= X_i - \gamma \theta_{,i} \;, & Y_i^* &= Y_i \;, & \underline{x} &\in V \;, \\ p_i^* &= p_i + \gamma \theta n_i \;, & m_i^* &= m_i \;, & \underline{x} &\in A_\sigma \;, \\ u_i^* &= u_i \;, & \varphi_i^* &= \varphi_i \;, & \underline{x} &\in A_u \;. \end{aligned} \tag{4.3.13}$$

Relations (4.3.13) represent the body forces analogy by means of which each steady-state problem can be reduced to the isothermal problem of the theory of asymmetric thermoelasticity.

4. 4 The Reciprocity Theorem.

Let us consider two systems of causes and effects acting on an elastic body of volume V bounded by the surface A .

We assign to the first group of causes the body forces $\underline{X}$, body couples $\underline{Y}$, heat sources Q , loadings $\underline{p}$ and $\underline{m}$ on the surface and the heating of this surface (the prescribed temperature or the heat flux on the surface A). The effects are: the components of the displacement vector $\underline{u}$, of the rotation vector $\underline{\varphi}$ and temperature ϑ .

The second set of causes and effects will be distinguished from the first one by a prime. In the sequel we assume the initial conditions to be homogeneous.

Let us apply the one-sided Laplace-transformation to the constitutive equations. We obtain the following relations

$$\bar{\sigma}_{ji} = (\mu+\alpha)\bar{\gamma}_{ji} + (\mu-\alpha)\bar{\gamma}_{ij} + (\lambda\bar{\gamma}_{kk} - \nu\bar{\vartheta})\delta_{ji}, \qquad (4.4.1)$$

$$\bar{\mu}_{ji} = (\gamma+\varepsilon)\bar{\varkappa}_{ji} + (\gamma-\varepsilon)\bar{\varkappa}_{ij} + \beta\bar{\varkappa}_{kk}\delta_{ji} , \qquad (4.4.2)$$

where

$$\sigma_{ji}(\underline{x},p) = \mathcal{L}(\sigma_{ji}(\underline{x},t)) = \int_0^\infty \sigma_{ji}(\underline{x},t)e^{-pt}dt , \quad \text{e.c.t.}$$

and similar relations for $\bar{\sigma}'_{ji}$ and $\bar{\mu}'_{ji}$. It may easily be seen that the following identity holds

$$\bar{\sigma}_{ji}\bar{\gamma}'_{ji} + \bar{\mu}_{ji}\bar{\varkappa}'_{ji} + \nu\bar{\vartheta}\bar{\gamma}'_{kk} = \bar{\sigma}'_{ji}\bar{\gamma}_{ji} + \bar{\mu}'_{ji}\bar{\varkappa}_{ji} + \nu\bar{\vartheta}'\bar{\gamma}_{kk} . \tag{4.4.3}$$

By integrating relation (4.4.3) over the volume V, we get

$$\int(\bar{\sigma}_{ji}\bar{\gamma}'_{ji} + \bar{\mu}_{ji}\bar{\varkappa}'_{ji} - \bar{\sigma}'_{ji}\bar{\gamma}_{ji} - \bar{\mu}'_{ji}\bar{\varkappa}_{ji})dV = \nu\int(\bar{\vartheta}'\bar{\gamma}_{kk} - \bar{\vartheta}\bar{\gamma}'_{kk})dV . \tag{4.4.4}$$

Now, let us perform the Laplace-transformation on the equations of motion

$$\bar{\sigma}_{ji,j} + \bar{X}_i = p^2\varrho\bar{u}_i , \qquad \epsilon_{ijk}\bar{\sigma}_{jk} + \bar{\mu}_{ji,j} + \bar{Y}_i = p^2 \mathfrak{J}\bar{\varphi}_i , \tag{4.4.5}$$

as well as on the equations of motion for the state marked with primes. We assume that the initial conditions of the equations of motion are homogeneous. By taking advantage of relations (4.4.5), we reduce Eq.(4.4.4) to the form

$$\int_V(\bar{X}_i\bar{u}'_i + \bar{Y}_i\bar{\varphi}'_i)dV + \int_A(\bar{p}_i\bar{u}'_i + \bar{m}_i\bar{\varphi}'_i)dA + \nu\int_V\bar{\vartheta}\bar{\gamma}'_{kk}dV =$$

$$= \int_V(\bar{X}'_i\bar{u}_i + \bar{Y}'_i\bar{\varphi}_i)dV + \int_A(\bar{p}'_i\bar{u}_i + \bar{m}'_i\bar{\varphi}_i)dA + \nu\int_V\bar{\vartheta}'\bar{\gamma}_{kk}dV . \tag{4.4.6}$$

This is the first part of the reciprocity theorem. The second part is obtained by taking account of the heat conduction equations for both sets.

We apply the Laplace-transformation to these equations assuming that the initial conditions are homogeneous

$$\bar{\vartheta}_{,jj} - \frac{p}{\varkappa}\bar{\vartheta} - \eta p\bar{\gamma}_{kk} = -\frac{\bar{Q}}{\varkappa} \tag{4.4.7}$$

$$\bar{\vartheta}'_{,jj} - \frac{p}{\varkappa}\bar{\vartheta}' - \eta p \bar{\gamma}'_{kk} = -\frac{\bar{Q}'}{\varkappa} .$$

By multiplying the first of Eqs.(4.4.7) by $\bar{\vartheta}'$, the second by $\bar{\vartheta}$, substracting one from another, integrating over the region V and making use of Green's transformation theorem, we get

$$p\eta \int_V (\bar{\gamma}_{kk}\bar{\vartheta}' - \bar{\gamma}'_{kk}\bar{\vartheta})dV + \frac{1}{\varkappa}\int_V (\bar{Q}'\bar{\vartheta} - \bar{Q}\bar{\vartheta}')dV - \\ - \int_A (\bar{\vartheta}'\bar{\vartheta}_{,n} - \bar{\vartheta}\bar{\vartheta}'_{,n})dA = 0. \qquad (4.4.8)$$

By eliminating the common term from (4.4.6) and (4.4.8), we obtain the final form of the theorem on reciprocity

$$\frac{\eta\varkappa p}{\nu}\Big[\int_V (\bar{X}_i\bar{u}'_i - \bar{X}'_i\bar{u}_i + \bar{Y}_i\bar{\varphi}'_i - \bar{Y}'_i\bar{\varphi}_i)dV + \\ + \int_A (\bar{p}_i\bar{u}'_i - \bar{p}'_i\bar{u}_i + \bar{m}_i\bar{\varphi}'_i - \bar{m}'_i\bar{\varphi}_i)dA + \\ + \varkappa\int_A (\bar{\vartheta}\bar{\vartheta}'_{,n} - \bar{\vartheta}'\bar{\vartheta}_{,n})dA + \int_V (\bar{Q}'\bar{\vartheta} - \bar{Q}\bar{\vartheta}')dV = 0 . \qquad (4.4.9)$$

It is obvious that in this relation all causes and effects appear. By applying the inverse Laplace-transformation on Eq.(4.4.9), we have

$$\frac{\eta\varkappa}{\nu}\Big\{\int_V dV(\underline{x})\int_0^t \Big[X_i(\underline{x},t-\tau)\frac{\partial u'_i(\underline{x},\tau)}{\partial\tau} - X'_i(\underline{x},t-\tau)\frac{\partial u_i(\underline{x},\tau)}{\partial\tau} + \\ + Y_i(\underline{x},t-\tau)\frac{\partial\varphi'_i(\underline{x},\tau)}{\partial\tau} - Y'_i(\underline{x},t-\tau)\frac{\partial\varphi_i(\underline{x},\tau)}{\partial\tau}\Big]d\tau + \qquad (4.4.10)$$

$$+\int_A dA(\underline{x})\int_0^t\Big[p_i(\underline{x},t-\tau)\frac{\partial u_i'(\underline{x},\tau)}{\partial\tau}-p_i'(\underline{x},t-\tau)\frac{\partial u_i(\underline{x},\tau)}{\partial\tau}+$$
$$+m_i(\underline{x},t-\tau)\frac{\partial\varphi_i'(\underline{x},\tau)}{\partial\tau}-m_i'(\underline{x},t-\tau)\frac{\partial\varphi_i(\underline{x},\tau)}{\partial\tau}\Big]d\tau\Big\}+$$
$$+\varkappa\int_A dA(\underline{x})\int_0^t[\vartheta(\underline{x},t-\tau)\vartheta_{,n}'(\underline{x},\tau)-\vartheta'(\underline{x},t-\tau)\vartheta_{,n}(\underline{x},\tau)]d\tau+$$
$$+\int_V dV(\underline{x})\int_0^t[\vartheta(\underline{x},t-\tau)Q'(\underline{x},\tau)-\vartheta'(\underline{x},t-\tau)Q(\underline{x},\tau)]d\tau\ .$$

With $Y_i = Y_i' = 0$, $m_i = m_i' = 0$ the theorem on reciprocity (4.4.10) goes over into the theorem on reciprocity for an elastic medium without couple-stresses, given by Cazimir - Ionescu+). For static loads and for stationary heat flow we get the system of equations

$$\int_V (X_i u_i' + Y_i\varphi_i')dV + \int_A (p_i u_i' + m_i\varphi_i')dA + \nu\int_V \vartheta\gamma_{kk}'dV =$$

$$= \int_V (X_i' u_i + Y_i'\varphi_i)dV + \int_A (p_i' u_i + m_i'\varphi_i)dA + \nu\int_V \vartheta'\gamma_{kk}dV\ , \tag{4.4.11}$$

$$\int_V (Q'\vartheta - Q\vartheta')dV + \varkappa\int_A (\vartheta_{,n}'\vartheta - \vartheta_{,n}\vartheta')dA = 0\ . \tag{4.4.12}$$

In the reciprocity equation (4.4.11) temperatures ϑ and ϑ' are treated as known functions, obtained from the solu-

---

+) Ionescu - Cazimir, V.: Problem of linear coupled thermoelasticity. Reciprocal theorems for the dynamic problem of thermoelasticity. Bull. Acad. Polon. Sci., Ser. Sci. Techn. 12, 9, 1964.

tion of the heat conductivity equations

$$\vartheta_{,jj} = -\frac{Q}{\varkappa} \quad , \quad \vartheta'_{,jj} = -\frac{Q'}{\varkappa} \quad . \qquad (4.4.13)$$

Equation (4.4.12) may be treated as a theorem of reciprocity for the problem of heat conduction.

4. 5 Conclusions from the Theorem of Reciprocity.

Let us consider first a simply-connected body. We are going to determine the integral

$$I = \int_V \gamma_{kk} dV \quad , \qquad \gamma_{kk} = -\operatorname{div} \underline{u} \, . \qquad (4.5.1)$$

Here I denotes the increment of the volume due to its deformation.

In order to determine the integral $I = \int_V \gamma_{kk} dV$, we assume that the second system of causes refers to the overall unary tension in isothermic state. Consequently, we have to assume

$$X'_i = 0 \quad , \quad Y'_i = 0 \quad , \quad m'_i = 0 \quad , \quad \vartheta' = 0 \quad , \quad p'_i = 1\, n_i . \qquad (4.5.2)$$

Since in this case there is

$$\sigma'_{ij} = 1\, \delta_{ij} \quad , \quad \gamma'_{ji} = \frac{1}{3K} \delta_{ij} \quad , \quad \varkappa'_{ji} = 0 \quad , \quad \mu'_{ji} = 0, \qquad (4.5.3)$$

we have

$$u'_i = \frac{x_i}{3K} \quad , \quad K = \lambda + \frac{2}{3}\mu \quad . \qquad (4.5.4)$$

From Eq.(4.4.11) we obtain

$$(4.5.5)\qquad \int_A n_i u_i \, dA = \frac{1}{3K}\left\{\int_A p_i x_i \, dA + \int_V X_i x_i \, dV\right\} + 3\alpha_t \int_V \vartheta \, dV .$$

As

$$\int_A n_i u_i \, dA = \int_V u_{k,k} \, dV = I$$

we get

$$(4.5.6)\qquad I = \int_V \gamma_{kk} \, dV = \frac{1}{3K}\left\{\int_A p_i x_i \, dA + \int_V X_i x_i \, dV\right\} + 3\alpha_t \int_V \vartheta \, dV.$$

The changes in the volume of the body depend here solely on the loadings p_i, body forces X_i and temperature ϑ. The moments m_i and body couples Y_i do not exert any influence on the changes of the volume of the body.

The changes of the volume of the body may be described in a particularly simple form if no external forces are acting. In this case there is

$$(4.5.7)\qquad I = 3\alpha \int_V \vartheta \, dV .$$

Let us remark that the increase of the volume of the body depends only on the distribution of heat in the body and on the coefficient of termal dilatation.

Formulae (4.5.6) and (4.5.7) are identical with those known from the classical theory of thermoelasticity+). Let

+) Nowacki, W.: Thermal stresses in anisotropic bodies I . (In Polish), Arch. Mech. Stos. 6, 3, (1955), 285.

us consider now the integral $\int_V \sigma_{kk} dV$. In virtue of constitutive equations we have

$$\int_V \sigma_{kk} dV = 3K \int_V \gamma_{kk} dV - 9K\alpha_t \int_V \vartheta dV . \tag{4.5.8}$$

By taking into account (4.5.6), we get finally

$$\int_V \sigma_{kk} dV = \int_A p_i x_i dA + \int_V X_i x_i dV . \tag{4.5.9}$$

If no external loadings are acting (the body being, however, heated) there is

$$\int_V \sigma_{kk} dV = 0 . \tag{4.5.10}$$

Eq.(4.5.10) is identical with Hieke's formula+) in classical thermoelasticity.

Let us consider, second, an infinite medium. Let a concentrated and instantaneous force $X_i = \delta(\underline{x} - \underline{\xi})\delta(t)\delta_{ik}$, directed along the axis x_k, be acting at the point $\underline{\xi}$ of the medium. Denote by $U_i^{(k)}(\underline{x}, \underline{\xi}, t)$ the displacement caused by this force. Furthermore, let a concentrated and instantaneous force $X_i' = \delta(\underline{x} - \underline{\eta})\delta(t)\delta_{ij}$, directed along the axis x_j , be acting at the point $\underline{\eta}$. Denote the displacement caused by this force by $U_i^{(j)}(\underline{x}, \underline{\xi}, t)$. From the theorem of reciprocity, formulated

---

+) Hieke,M.: Über ein ebenes Distorsions problem. 35, Z.A.M.M., (1955), 54.

for an infinite region, we have

$$\int_V dV(\underline{x})\int_0^t d\tau\Big[\delta(\underline{x}-\underline{\xi})\delta(t-\tau)\delta_{ik}\frac{\partial U_i^{(j)}(\underline{x},\underline{\eta},\tau)}{\partial\tau} -$$

(4.5.11)
$$-\delta(\underline{x}-\underline{\eta})\delta(t-\tau)\delta_{ij}\frac{\partial U_i^{(k)}(\underline{x},\underline{\xi},\tau)}{\partial\tau}\Big] = 0$$

and hence

$$\dot{U}_k^{(j)}(\underline{\xi},\underline{\eta},t) = \dot{U}_j^{(k)}(\underline{\eta},\underline{\xi},t) .$$

After integration with respect to time, we finally get

$$U_k^{(j)}(\underline{\xi},\underline{\eta},t) = U_j^{(k)}(\underline{\eta},\underline{\xi},t) . \tag{4.5.12}$$

Let a concentrated and instantaneous force $X_i = \delta(\underline{x}-\underline{\xi})\delta(t)\delta_{ik}$ act at the point $\underline{\xi}$ of an infinite medium, and a concentrated and instantaneous source of heat $Q' = \delta(\underline{x}-\underline{\eta})\delta(t)$ act at the point $\underline{\eta}$. Denote by $\Theta^{(k)}(\underline{x},\underline{\xi},t)$ the temperature caused by the action of the force X_i, and by $U_i(\underline{x},\underline{\eta},t)$ the displacement caused by the action of the source Q. From the theorem of reciprocity we obtain the following relation

$$\int_V dV(\underline{x})\int_0^t d\tau\Big[\delta(\underline{x}-\underline{\eta})\delta(t-\tau)\frac{\partial\Theta(\underline{x},\underline{\xi},\tau)}{\partial\tau} +$$

(4.5.13)
$$+\frac{\eta\varkappa}{\nu}\delta(\underline{x}-\underline{\xi})\delta(t-\tau)\delta_{ik}\frac{\partial U_i(\underline{x},\underline{\eta},\tau)}{\partial\tau}\Big] = 0 ,$$

wherefrom

$$\Theta^{(k)}(\underline{\eta},\underline{\xi},t) = -\frac{\eta\varkappa}{\nu}\frac{\partial U_k(\underline{\xi},\underline{\eta},t)}{\partial t} . \tag{4.5.14}$$

Let a concentrated and instantaneous force $X_i = \delta(\underline{x}-\underline{\xi})\delta(t)\delta_{ik}$ act at the point $\underline{\xi}$ of infinite medium, and the concentrated and instantaneous body couple $Y_i' = \delta(\underline{x} - \underline{\eta})\delta(t)\delta_{ij}$ act at the point $\underline{\eta}$. Denote by $\Omega_i^{(k)}(\underline{x}, \underline{\xi}, t)$ the rotation vector caused by the action of force X_i and by $V_i^{(j)}(\underline{x}, \underline{\eta}, t)$ the displacement caused by the body couple Y_i'. From the theorem of reciprocity we get:

$$V_k^{(j)}(\underline{\xi}, \underline{\eta}, t) = \Omega_j^{(k)}(\underline{\eta}, \underline{\xi}, t) \ . \qquad (4.5.15)$$

Finally, let a body couple $Y_i = \delta(\underline{x} - \underline{\xi})\delta(t)\delta_{ik}$ act at the point $\underline{\xi}$, and a source of heat $Q' = \delta(\underline{x} - \underline{\xi})\delta(t)$ act at the point $\underline{\eta}$. Denote the temperature caused by the action of body couple by $\vartheta^{(k)}(\underline{x}, \underline{\xi}, t)$ and the rotation vector caused by the action of the source Q' by $\Phi_i(\underline{x}, \underline{\eta}, t)$. From the theorem of reciprocity we obtain the following relation

$$\vartheta^{(k)}(\underline{\eta}, \underline{\xi}, t) = -\frac{\eta \varkappa}{\gamma} \frac{\partial \Phi_k(\underline{\xi}, \underline{\eta}, t)}{\partial t} \ . \qquad (4.5.16)$$

It can be shown that the relations (4.5.12) (4.5.14) (4.5.16) hold for a finite body at homogeneous boundary conditions.

Let us consider a finite body V and assume that the causes which set the medium in motion are defined by the boundary conditions. We wish to find expressions for the displacements $\underline{u}$ rotation vector $\underline{\varphi}$ and temperature θ at an internal point $\underline{x} \in V$ by means of integrals on the surface A bounding the region V. These functions should satisfy the equations of motion, the extended equation of heat conduction and the boundary conditions. When deriving the formulae for the functions $u_i(\underline{x},t)$ and $\varphi_i(\underline{x},t)$, $\vartheta(\underline{x},t)$, we shall use the theorem of reciprocity. Assume, first, that quantities marked with primes refer to displacements $u_i' = U_i^{(k)}(\underline{x},\underline{\xi},t)$, rotation vector $\varphi_i' = \Omega_i^{(k)}(\underline{x},\underline{\xi},t)$ and temperature $\vartheta' = \Theta^{(k)}(\underline{x},\underline{\xi},t)$, caused in an infinite medium by a concentrated and instantaneous force $X_i' = \delta(\underline{x}-\underline{\xi})\delta(t)\delta_{ik}$, applied at the point $\underline{\xi}$ and directed along the axis x_k. By assuming absence of body forces ($X_i = 0$), body couples ($Y_i = Y_i' = 0$) and heat sources ($Q = Q' = 0$), we obtain from the theorem of reciprocity the following expression

$$\begin{aligned}\dot{u}_k(x,t) = \int_A dA(\underline{\xi})\int_0^t d\tau \Big\{ & p_i(\underline{\xi},\tau)\frac{\partial U_i^{(k)}(\underline{\xi},\underline{x},t-\tau)}{\partial\tau} - \\ & - p_i^{(k)}(\underline{\xi},\underline{x},t-\tau)\frac{\partial u_i(\underline{\xi},\tau)}{\partial\tau} + m_i(\underline{\xi},t-\tau)\frac{\partial\Omega_i^{(k)}(\underline{\xi},\underline{x},\tau)}{\partial\tau} - \\ & - m_i^{(k)}(\underline{\xi},\underline{x},t-\tau)\frac{\partial\varphi_i(\underline{\xi},\tau)}{\partial\tau} + \frac{\gamma}{\eta}\big[\vartheta(\underline{\xi},t-\tau)\Theta_{,n}^{(k)}(\underline{\xi},\underline{x},\tau) - \\ & - \Theta^{(k)}(\underline{\xi},\underline{x},t-\tau)\vartheta_{,n}(\underline{\xi},\tau)\big]\Big\}, \quad \underline{x}\in V, \ \underline{\xi}\in A. \end{aligned} \tag{4.5.17}$$

Here we have introduced the notations

$$p_i^{(k)} = \sigma_{ji}^{(k)}(\underline{x},\underline{\xi},t)n_j(\underline{x}), \quad m_i^{(k)}(\underline{x},\underline{\xi},t) = \mu_{ji}^{(k)}(\underline{x},\underline{\xi},t)n_j(\underline{x}), \quad \underline{x}\in A$$

where by $\sigma_{ji}^{(k)}$ we understand stresses and by $\mu_{ji}^{(k)}$ couple-stresses caused by a concentrated force $X_i' = \delta(\underline{x}-\underline{\xi})\delta(t)\delta_{ik}$.

Integration operations in the surface integrals are to be carried out with respect to the variable $\underline{\xi}$. Formula (4.5.17) gives us the relation between the function $\dot{u}_k(\underline{x},t)$,

$x \in V$, $t>0$ and functions u_i, p_i, m_i, φ_i, ϑ, $\vartheta_{,n}$ on the surface A.

Now, let us assume, in the system with "primes", a concentrated and instantaneous body couple $Y_i' = \delta(\underline{x}-\underline{\xi})\delta(t)\delta_{ik}$ acting along the axis x_k. In an infinite medium the body couple will cause the displacement $u_i' = V_i^{(k)}(\underline{x},\underline{\xi},t)$, vector $\varphi_i' = \Lambda_i^{(k)}(\underline{x},\underline{\xi},t)$ and temperature $\vartheta' = \vartheta^{(k)}(\underline{x},\underline{\xi},t)$. From the theorem of reciprocity, at $X_i = X_i' = 0$, $Y_i = 0$, $Q = Q' = 0$ we obtain the following formula:

$$\begin{aligned}\dot{\varphi}_k(\underline{x},t) = \int_A dA(\underline{\xi}) \int_0^t d\tau \Big\{ & p_i(\underline{\xi},t-\tau)\frac{\partial V_i^{(k)}(\underline{\xi},\underline{x},\tau)}{\partial\tau} - \\ & -\hat{p}_i^{(k)}(\underline{\xi},\underline{x},t-\tau)\frac{\partial u_i(\underline{\xi},\tau)}{\partial\tau} + m_i(\underline{\xi},t-\tau)\frac{\partial \Lambda_i^{(k)}(\underline{\xi},\underline{x},\tau)}{\partial\tau} - \\ & -\hat{m}_i^{(k)}(\underline{\xi},\underline{x},t-\tau)\frac{\partial\varphi_i(\underline{\xi},\tau)}{\partial\tau} + \frac{\nu}{\eta}[\vartheta(\underline{\xi},t-\tau)\vartheta_{,n}^{(k)}(\underline{\xi},\underline{x},\tau) - \\ & - \vartheta^{(k)}(\underline{\xi},\underline{x},t-\tau)\vartheta_{,n}(\underline{\xi},\tau)]\Big\}, \quad \underline{x}\in V,\ \underline{\xi}\in A. \end{aligned} \tag{4.5.18}$$

Here

$$\hat{p}_i^{(k)} = \hat{\sigma}_{ji}^{(k)}(\underline{x},\underline{\xi},t)n_j(\underline{x}), \quad \hat{m}_i^{(k)} = \hat{\mu}_{ji}^{(k)}(\underline{x},\underline{\xi},t)n_j(\underline{x}).$$

We denote by $\hat{\sigma}_{jk}^{(k)}$ and by $\hat{\mu}_{ji}^{(k)}$ the force-stress tensor and the couple-stress tensor, respectively. Also the function $\dot{\varphi}_k(\underline{x},t)$, $x \in V$, $t>0$ is expressed here by the functions u_i, p_i, φ_i, m_i, ϑ, $\vartheta_{,n}$ on the surface A. Now, let the system with "primes" in an infinite medium be limited to the action of a concentrated and instantaneous heat source $Q' = \delta(\underline{x}-\underline{\xi})\delta(t)$ causing displacements $u_i' = U_i(\underline{x},\underline{\xi},t)$ rotation vector $\varphi_i' = \Phi_i(\underline{x},\underline{\xi},t)$ and temperature $\vartheta' = \Theta(\underline{x},\underline{\xi},t)$. From the theorem of reciprocity, by assuming that $X_i = X_i' = 0$, $Y_i = Y_i' = 0$, $Q = 0$, we get the temperature at the point

$$\vartheta(\underline{x},t) = \varkappa \int_A dA(\underline{\xi}) \int_0^t d\tau \Big\{ \vartheta_{,n}(\underline{\xi},\tau)\Theta(\underline{\xi},\underline{x},t-\tau) -$$

$$- \vartheta(\underline{\xi},t-\tau)\Theta_{,n}(\underline{\xi},\underline{x},\tau) - \frac{\eta}{\nu}\Big[p_i(\underline{\xi},t-\tau)\frac{\partial U_i(\underline{\xi},\underline{x},\tau)}{\partial\tau} -$$

$$- p_i^*(\underline{\xi},\underline{x},t-\tau)\frac{\partial u_i(\underline{\xi},\tau)}{\partial\tau} + m_i(\underline{\xi},t-\tau)\frac{\partial \Phi_i(\underline{\xi},\underline{x},\tau)}{\partial\tau} -$$

$$- m_i^*(\underline{\xi},\underline{x},t-\tau)\frac{\partial \varphi_i(\underline{\xi},\tau)}{\partial\tau}\Big]\Big\}, \quad \underline{x}\in V, \; \underline{\xi}\in A. \tag{4.5.19}$$

Here

$$p_i^*(\underline{x},t) = \sigma_{ji}^*(\underline{x},\underline{\xi},t)n_j(\underline{x}) \quad , \quad m_i^*(\underline{x},t) = \mu_{ji}^*(\underline{x},\underline{\xi},t)n_j(\underline{x}) .$$

We denote by σ_{ji}^* and by μ_{ji}^* the force-stress tensor and couple-stress tensor caused by the action of an instantaneous and concentrated heat source Q'.

Formulae (4.5.17) ÷ (4.5.19) may be treated as an extension of Somigliana's formulae, to the problems of thermoelasticity. Some simplifications of these formulae can be obtained by taking into account the reciprocity relations (4.5.14) ÷ (4.5.16).

If Green's functions $U_i^{(k)}$, $\Omega_i^{(k)}$, $\Theta^{(k)}$,... etc. are selected in such a way as to satisfy on the surface A the homogeneous boundary conditions for displacements, rotation vec tor and temperature, then, the formulae (4.5.17)÷(4.5.19) will yield the solution of the first boundary problem, if functions u_i, φ_i and ϑ are given at the boundary.

Similarly, if we select Green's functions $U_i^{(k)}$, $\Omega_i^{(k)}$, $\Theta^{(k)}$,.... etc. in such a way that the boundary is free from loads and temperature, then the formulae (4.5.17)÷(4.5.19) will yield the solution of the second boundary value problem, when on A loads p_i, m_i and temperature ϑ are given.

4. 6 Generalized Maysel's Formulae.

Let us consider a micropolar elastic body subjected to heating. The displacements $u(\underline{x},t)$, rotations $\varphi(\underline{x},t)$ and the temperature as well have to verify the following system of differential equations

$$\begin{aligned} &L(\underline{u}) + M(\underline{\varphi}) + N(\vartheta) = 0\,, \\ &M(\underline{u}) + K(\underline{\varphi}) = 0\,, \\ &D(\vartheta) + \frac{Q}{\varkappa} = 0\,. \end{aligned} \tag{4.6.1}$$

The following notations have been introduced in Eq.(4.6.1)

$$L(\underline{u}) = \Box_2 \underline{u} + (\lambda + \mu - \alpha)\,\mathrm{grad\,div}\,\underline{u}\,, \quad M(\underline{u}) = 2\alpha\,\mathrm{rot}\,\underline{u}\,,$$

$$K(\underline{\varphi}) = \Box_4 \underline{\varphi} + (\beta + \gamma - \varepsilon)\,\mathrm{grad\,div}\,\underline{\varphi}\,,$$

$$N(\vartheta) = -\nu\,\mathrm{grad}\,\vartheta\,, \qquad D(\vartheta) = \left(\nabla^2 - \frac{1}{\varkappa}\partial_t\right)\vartheta\,.$$

Let us assume that on the surface A , bounding the body, the following homogeneous mixed boundary conditions are prescribed:

$$\begin{aligned} &\underline{u}=0\,, \quad \underline{\varphi}=0\,, \quad \vartheta_{,n}=0\,, && \underline{x}\in A_u\,, \\ &\underline{p}=0\,, \quad \underline{m}=0 \quad \vartheta=0\,, && \underline{x}\in A_\sigma\,. \end{aligned} \tag{4.6.2}$$

We assume the initial conditions to be homogeneous.

In the sequel we shall make use of the theorem on the reciprocity of works. By considering two systems of causes and effects (the second one will be marked with "primes"), we obtain for the problem of non-coupled thermoelasticity the following equation, the initial conditions of the functions $\underline{u}$, $\underline{u}'$, $\underline{\varphi}$, $\underline{\varphi}'$ being assumed homogeneous

$$\begin{aligned} &\int_V (X_i * u_i' + Y_i * \varphi_i')dV + \int_A (p_i * u_i' + m_i * \varphi_i')dA + \nu \int_V \vartheta * \gamma_{kk}' dV = \\ &= \int_V (X_i' * u_i + Y_i' * \varphi_i)dV + \int_A (p_i' * u_i + m_i' * \varphi_i)dA + \nu \int_V \vartheta' * \gamma_{kk} dV, \end{aligned} \tag{4.6.3}$$

where

$$X_i * u_i' = \int_0^t X_i(\underline{x}, t-\tau)u_i'(\underline{x},\tau)d\tau = \int_0^t X_i(\underline{x},\tau)u_i'(\underline{x}, t-\tau)d\tau \ ,$$

and so on.

Now, let us assume that an instantaneous concentrated unitary force $\underline{X}' = \delta(\underline{x}-\underline{\xi})\delta(t)e_j$ - directed in parallel to the x_j-axis - is acting at the point $\underline{\xi}\in V$ of the body in isothermal state $(\vartheta'=0)$. The action of this force will induce in the body displacements $\underline{u}' = \underline{U}^{(j)}(\underline{x},\underline{\xi},t)$ and rotations $\underline{\varphi}' = \underline{\Phi}^{(j)}(\underline{x},\underline{\xi},t)$.

These functions have to verify the following differential equations

$$\begin{aligned} &L(\underline{U}^{(j)}) + M(\underline{\Phi}^{(j)}) + \delta(\underline{x} - \underline{\xi})\delta(t)\underline{e}_j = 0 , \\ &M(\underline{U}^{(j)}) + K(\underline{\Phi}^{(j)}) = 0 . \end{aligned} \qquad (4.6.4)$$

The boundary conditions are assumed to be homogeneous, i.e. there is

$$\begin{aligned} &\underline{U}^{(j)} = 0 , \quad \underline{\Phi}^{(j)} = 0 , \qquad \underline{x} \in A_u , \\ &\underline{p}^{(j)} = 0 , \quad \underline{m}^{(j)} = 0 , \qquad \underline{x} \in A_\sigma . \end{aligned}$$

Here, $\underline{p}^{(j)}$ denotes the main stress vector, while $\underline{m}^{(j)}$ stands for the main couple-stress vector on the surface A_σ .

We shall apply the theorem of the reciprocity of works, Eq.(4.6.3), to the two systems of causes and effects considered in this paper. As a result we obtain the following formula

$$u_j(\underline{\xi}, t) = \nu \int_V dV(\underline{x}) \int_0^t \vartheta(\underline{x}, t-\tau) U^{(j)}_{k,k}(\underline{x}, \underline{\xi}, \tau) d\tau. \qquad (4.6.5)$$

Consider another system of loadings, that is marked by "primes". An instantaneous concentrated unitary body couple $\underline{Y}' = \delta(\underline{x} - \underline{\xi})\delta(t)\underline{e}_j$ is supposed to act parallel to the x_j -axis at the point $\underline{\xi} \in V$ of the body in isothermal state $(\vartheta' = 0)$. We denote the displacements and rotations induced by this action by $\underline{u}' = \underline{V}^{(j)}(\underline{x}, \underline{\xi}, t)$ and $\underline{\varphi}' = \underline{\Gamma}^{(j)}(\underline{x}, \underline{\xi}, t)$, respectively .

These functions have to verify the differential equations of micropolar thermoelasticity

$$\text{(4.6.6)}\qquad \begin{aligned} &L(V^{(j)}) + M(\Gamma^{(j)}) = 0 \ , \\ &M(V^{(j)}) + K(\Gamma^{(j)}) + \delta(\underline{x}-\underline{\xi})\delta(t)\underline{e}_j = 0 \ , \end{aligned}$$

by assuming the homogeneity of initial and boundary (on the surface A_u and A_σ) conditions.

By making use of the theorem on reciprocity, Eq. (4.6.3), we obtain

$$\text{(4.6.7)}\qquad \varphi_j(\underline{\xi},t) = \nu\int_V dV(\underline{x})\int_0^t \vartheta(\underline{x},t-\tau)V^{(j)}_{k,k}(\underline{x},\underline{\xi},\tau)d\tau .$$

Formulae (4.6.5) and (4.6.7) represent the generalization of Maysel's formulae on the problems of micropolar thermoelasticity.

We may rewrite the expressions (4.6.5) and (4.6.7) in a somewhat modified form

$$\text{(4.6.8)}\qquad \begin{aligned} u_j(\underline{\xi},t) &= \nu\int_V dV(\underline{x})\int_0^t \vartheta(\underline{x},t-\tau)W_j(\underline{\xi},\underline{x},\tau)d\tau \ , \\ \varphi_j(\underline{\xi},t) &= \nu\int_V dV(\underline{x})\int_0^t \vartheta(\underline{x},t-\tau)\Omega_j(\underline{\xi},\underline{x},\tau)d\tau \ . \end{aligned}$$

The term $W_j(\underline{x},\underline{\xi},t)$ stands here for the displacement, while $\Omega_j(\underline{x},\underline{\xi},t)$ for the rotation induced at the point $\underline{x}$ owing to the action of the pressure center situated at the point $\underline{\xi}$. The relations

$$\text{(4.6.9)}\qquad W_j(\underline{\xi},\underline{x},t) = U^{(j)}_{k,k}(\underline{x},\underline{\xi},t)\ ,\quad \Omega_j(\underline{\xi},\underline{x},t) = V^{(j)}_{k,k}(\underline{x},\underline{\xi},t),$$

are also the consequence of the theorem on reciprocity,(4.6.3). For the steady-state problems of thermoelasticity the theorem on reciprocity takes the following form:

$$\int_V (X_i u_i' + Y_i \varphi_i') dV + \int_A (p_i u_i' + m_i \varphi') dA + \nu \int_V \vartheta \gamma_{kk}' dV =$$
$$= \int_V (X_i' u_i + Y_i' \varphi_i) dV + \int_A (p_i' u_i + m_i' \varphi_i) dA + \nu \int_V \vartheta' \gamma_{kk} dV. \quad (4.6.10)$$

By proceeding similarly as in the dynamic problem, we get

$$\begin{aligned} u_j(\underline{\xi}) &= \nu \int_V \vartheta(\underline{x}) U_{k,k}^{(j)}(\underline{x}, \underline{\xi}) dV(\underline{x}) \\ \varphi_j(\underline{\xi}) &= \nu \int_V \vartheta(\underline{x}) V_{k,k}(\underline{x}, \underline{\xi}) dV(\underline{x}) \end{aligned} \quad (4.6.11)$$

The expressions (4.6.5) ÷ (4.6.7) may be written in a particularly simple form in the case of an infinite space. By solving the systems of Eqs.(4.6.4) and (4.6.6), we obtain

$$u_j(\underline{\xi}, t) = \frac{\nu}{4\pi(\lambda + 2\mu)} \frac{\partial}{\partial x_j} \int_V \frac{\vartheta(\underline{x}, t - R/c_1)}{R(\underline{x}, \underline{\xi})} dV(\underline{x}),$$
$$\varphi_j(\underline{\xi}, t) = 0 . \quad (4.6.12)$$

The latter result is due to the fact that $V_{k,k}^{(j)} = 0$. For the steady-state problem there is

$$\begin{aligned} u_j(\underline{\xi}) &= \frac{\nu}{4\pi(\lambda + 2\mu)} \frac{\partial}{\partial x_j} \int_V \frac{\vartheta(\underline{x}) dV(\underline{x})}{R(\underline{x}, \underline{\xi})} , \\ \varphi_j(\underline{\xi}) &= 0 . \end{aligned} \quad (4.6.13)$$

The above solutions are identical with those obtained in classical thermoelasticity.

4. 7 Problems of Thermoelasticity with Axial Symmetry and Symmetry with Respect to the Point.

Let us consider the axi-symmetric problem, the temperature being assumed to depend on the variables r, z and t. Within the system of cylindrical coordinates (r,ϑ,z) - under the assumption that all the causes of effects are independent of the angle ϑ - we obtain the following equations of thermoelasticity:

$$(4.7.1)\quad \begin{aligned} &(\mu+\alpha)\left(\nabla^2-\frac{1}{r^2}\right)u_r+(\lambda+\mu-\alpha)\frac{\partial e}{\partial r}-2\alpha\frac{\partial\varphi_\vartheta}{\partial z}=\rho\ddot{u}_r+\nu\frac{\partial\theta}{\partial r},\\ &(\mu+\alpha)\nabla^2u_z+(\lambda+\mu-\alpha)\frac{\partial e}{\partial z}+\frac{2\alpha}{r}\frac{\partial}{\partial r}(r\varphi_\vartheta)=\rho\ddot{u}_z+\nu\frac{\partial\theta}{\partial z},\\ &(\gamma+\varepsilon)\left(\nabla^2-\frac{1}{r^2}\right)\varphi_\vartheta-4\alpha\varphi_\vartheta+2\alpha\left(\frac{\partial u_r}{\partial z}-\frac{\partial u_z}{\partial r}\right)=J\ddot{\varphi}_\vartheta \end{aligned}$$

where

$$\nabla^2=\frac{\partial^2}{\partial r^2}+\frac{1}{r}\frac{\partial}{\partial r}+\frac{\partial^2}{\partial z^2},\qquad e=\frac{1}{r}\frac{\partial}{\partial r}(ru_r)+\frac{\partial u_z}{\partial z},\qquad r=(x_1^2+x_2^2)^{1/2}.$$

The case of temperature $\theta(r,t)$ distribution, where $u_z=0$ and $\varphi_\vartheta=0$, is of particular interest. By neglecting the derivatives with respect to z, we obtain the following form of Eq.(4.7.1)

$$(4.7.2)\quad \begin{aligned} &(\lambda+2\mu)\left(\nabla_r^2-\frac{1}{r^2}\right)u_r=\rho\ddot{u}_r+\nu\frac{\partial\theta}{\partial r},\\ &\nabla^2=\frac{\partial^2}{\partial r^2}+\frac{1}{r}\frac{\partial}{\partial r}. \end{aligned}$$

The above equation is identical in the form with that of classical thermoelasticity.

Maysel's formulae for this case may be presented in the

form [+)]

$$u_r(r,t)=\frac{\nu}{r}\int_a^b r'dr'\int_0^t \theta(r',t-\tau)\hat{e}(r',r,\tau)d\tau\ ,\quad \varphi_\vartheta=0\ ,\quad u_z=0, \tag{4.7.3}$$

or else

$$u_r(r,t)=\frac{\nu}{r}\int_a^b r'dr'\int_0^t \theta(r',t-\tau)W_r(r,r',\tau)d\tau\ ,\quad \varphi_\vartheta=0\ ,\quad u_z=0. \tag{4.7.4}$$

Formulae (4.7.3) and (4.7.4) refer to an infinite hollow cylinder with inner radius a and outer radius b. Here, the term $U_r(r',r,t)$ represents the displacement on the surface of the cylinder $r=$const., due to the action of instantaneous radial forces distributed uniformly on the surface of the cylinder $r=$ const. The quantity $\hat{e}=\frac{U_r}{r'}+\frac{\partial U_r}{\partial r'}$ is the dilatation due to the action of radial forces on the surface $r=$ const.The function $W_r(r,r',t)$ should be regarded as the displacement on the surface $r=$ const., induced by the action of instantaneous pressure centers distributed uniformly on the surface of the cylinder $r=$ const.

Formulae (4.7.3) and (4.7.4) are identical with those obtained in classical thermoelasticity.

We shall, now, consider the axi-symmetric thermoelastic problem. To begin with, we shall rewrite the fundamental equations so as to fit the spherical coordinate system (R,ϑ,η). By assuming axial symmetry with respect to x_3, we obtain the

---

+) Maysel, V.M.: Temperature problems of the theory of elasticity. (in Russian), Kiev, 1951.

following system of equations describing the displacements $\underline{u} \equiv (u_R, u_\vartheta, 0)$ and the rotations $\underline{\varphi} \equiv (0, 0, \varphi_\eta)$:

$$(\mu+\alpha)\left\{\nabla^2 u_R - \frac{2}{R^2}\left[u_R + \frac{1}{\sin\vartheta}\frac{\partial}{\partial\vartheta}(u_\vartheta \sin\vartheta)\right]\right\} +$$

$$+ (\lambda+\mu-\alpha)\frac{\partial e}{\partial R} + \frac{2\alpha}{R}\frac{1}{\sin\vartheta}\frac{\partial}{\partial\vartheta}(\varphi_\vartheta \sin\vartheta) - \rho\ddot{u}_R = \nu\frac{\partial\theta}{\partial R},$$

$$(\mu+\alpha)\left\{\nabla^2 u_\vartheta - \frac{2}{R^2}\left[\frac{\partial u_R}{\partial\vartheta} - \frac{u_\vartheta}{2\sin^2\vartheta}\right]\right\} + (\lambda+\mu-\alpha)\frac{1}{R}\frac{\partial e}{\partial R} +$$

$$\text{(4.7.5)} \qquad - \frac{2\alpha}{R}\frac{\partial}{\partial R}(R\varphi_\eta) - \rho\ddot{u}_\vartheta = \frac{\nu}{R}\frac{\partial\theta}{\partial\vartheta},$$

$$(\gamma+\varepsilon)\left\{\nabla^2\varphi_\eta - \frac{\varphi_\vartheta}{R^2\sin^2\vartheta}\right\} - 4\alpha\varphi_\eta +$$

$$- 2\alpha\left(\frac{1}{R}\frac{\partial u_R}{\partial\vartheta} - \frac{1}{R}\frac{\partial}{\partial R}(R u_\vartheta)\right) = \Im\ddot{\varphi}_\eta = 0 .$$

In the above equations the following notations have been used

$$\nabla^2 u_R = \frac{\partial^2 u_R}{\partial R^2} + \frac{2}{R^2}\frac{\partial u_R}{\partial R} + \frac{1}{R^2\sin^2\vartheta}\frac{\partial}{\partial\vartheta}\left(\sin\vartheta\frac{\partial u_R}{\partial\vartheta}\right),$$

$$e = \frac{1}{R^2}\frac{\partial}{\partial R}(u_R R^2) + \frac{1}{R\sin\vartheta}\frac{\partial}{\partial\vartheta}(u_\vartheta\sin\vartheta), \quad R = (x_1^2 + x_2^2 + x_3^2)^{1/2} .$$

In the case of symmetry with respect to the point, all the derivatives with respect to ϑ should be dropped and, moreover, $u_\vartheta = 0$ and $\varphi_\eta = 0$. Thus, what remains from the system of Eqs.(4.7.5) is the equation (4.7.5) which now reads as follows

$$(\lambda+2\mu)\left(\nabla_R^2 - \frac{1}{R^2}\right)u_R - \rho\ddot{u}_R = \nu\frac{\partial\theta}{\partial R},$$

$$\text{(4.7.6)} \qquad \nabla_R^2 = \frac{\partial^2}{\partial R^2} + \frac{2}{R}\frac{\partial}{\partial R} .$$

The above equation is identical in the form with that of classical thermoelasticity. The displacement $u_R(R,t)$ may be described by the formula

$$u_R(R,t)=\frac{\gamma}{R^2}\int_a^b R'^2 dR'\int_0^t \theta(R',t-\tau)\hat{e}(R',R)d\tau,\quad u_\vartheta=0,\ \varphi_\eta=0, \quad (4.7.7)$$

or

$$u_R(R,t)=\frac{\gamma}{R^2}\int_a^b R'^2 dR'\int_0^t \theta(R',t-\tau)W_R(R,R',\tau)d\tau,\quad u_\vartheta=0,\ \varphi_\eta=0. \quad (4.7.8)$$

These formulae refer to a hollow sphere with inner radius a and outer radius b. The function $\hat{e}$ represents the dilatation induced on the surface $R'=$ const. by the action of instantaneous radial forces distributed uniformly on the surface $R=$ const. The function $W_R(R,R',t)$ expresses the radial displacements on the surface $R=$ const., due to the action of instantaneous unitary pressure centers distributed uniformly on the surface $R'=$ const.

Our considerations presented here lead to a general conclusion: all problems - be they static, quasi-static or dynamic - characterized by the symmetry with respect to the point and referring to a hollow sphere have identical, in form, solutions for micropolar and Hooke's media. The same holds true for a full sphere with $a\to 0$, for an infinite space with a spherical cavity with $b\to\infty$ and for an infinite space with $a\to 0$, $b\to 0$.

Let us remark that we can argue quite similarly in undimensional problems (those of infinite space, half-space, elastic layer) where the temperature depends solely on the variables x_1 and t.

4. 8 The Plane Problem of Thermoelasticity.

In this Section we shall be concerned with plane state of strain induced in an elastic micropolar medium by the action of temperature.

We confine ourselves to the problem of stationary flow of heat.

By assuming that the displacements and rotations do not depend on the variable x_3, we have

(4.8.1) $$\underline{u} = (u_1, u_2, 0) \quad , \quad \underline{\varphi} = (0, 0, \varphi_3) \quad .$$

From the Section 3.4. we have the following compatibility relations

(4.8.2) $$\begin{aligned} &\partial_1^2 \gamma_{22} + \partial_2^2 \gamma_{11} = \partial_1 \partial_2 (\gamma_{12} + \gamma_{21}) \; , \\ &\partial_2^2 \gamma_{12} - \partial_1^2 \gamma_{21} = \partial_1 \partial_2 (\gamma_{22} - \gamma_{11}) - (\partial_1 \varkappa_{13} + \partial_2 \varkappa_{23}) \; , \\ &\partial_1 \varkappa_{23} - \partial_2 \varkappa_{13} = 0 \; . \end{aligned}$$

From the constitutive equations we get

(4.8.3) $$\begin{aligned} &\sigma_{ji} = (\mu + \alpha)\gamma_{ji} + (\mu - \alpha)\gamma_{ij} + (\lambda \gamma_{kk} - \nu \vartheta)\delta_{ji} \\ &\sigma_{33} = (\lambda \gamma_{kk} - \nu\vartheta), \; \mu_{j3} = (\gamma + \varepsilon)\varkappa_{j3}, \; \mu_{3j} = (\gamma - \varepsilon)\varkappa_{j3}, \; i,j = 1,2. \end{aligned}$$

Here $\gamma_{kk} = \gamma_{11} + \gamma_{22}$.

By solving Eqs.(4.8.3) for the strains γ_{ji} and

$\varkappa_{ji}$ and introducing the above relations into the compatibility relations (4.8.2), we arrive at the following three equations in stresses

$$\partial_2^2\sigma_{11} + \partial_1^2\sigma_{22} - \frac{\lambda}{2(\lambda+\mu)}\nabla_1^2(\sigma_{11}+\sigma_{22}) + \frac{\mu\nu}{\lambda+\mu}\nabla_1^2\theta = \partial_1\partial_2(\sigma_{12}+\sigma_{21})$$

$$(\partial_2^2 - \partial_1^2)(\sigma_{12}+\sigma_{21}) + \frac{\mu}{\alpha}\nabla_1^2(\sigma_{12}-\sigma_{21}) = 2\partial_1\partial_2(\sigma_{22}-\sigma_{11}) +$$

$$- \frac{4\mu}{\gamma+\varepsilon}(\partial_1\mu_{13} + \partial_2\mu_{23})$$

$$\partial_1\mu_{23} - \partial_2\mu_{13} = 0 \quad , \qquad \nabla_1^2 = \partial_1^2 + \partial_2^2 \ . \qquad (4.8.4)$$

We shall now introduce the function of stresses F and Ψ and connect them with the stresses by the following relations

$$\sigma_{11} = \partial_2^2 F - \partial_1\partial_2\Psi \ , \qquad \sigma_{22} = \partial_1^2 F + \partial_1\partial_2\Psi \ ,$$

$$\sigma_{12} = -\partial_1\partial_2 F - \partial_2^2\Psi \ , \qquad \sigma_{21} = -\partial_1\partial_2 F + \partial_1\Psi \qquad (4.8.5)$$

$$\mu_{13} = \partial_1\Psi \quad , \qquad \mu_{23} = \partial_2\Psi \ .$$

By substituting relations (4.8.4) into the equations of equilibrium

$$\partial_1\sigma_{11} + \partial_2\sigma_{21} = 0 \ ,$$

$$\partial_1\sigma_{12} + \partial_2\sigma_{22} = 0 \ , \qquad (4.8.6)$$

$$\sigma_{12} - \sigma_{21} + \partial_1\mu_{13} + \partial_2\mu_{23} = 0 \ ,$$

we see that the equations are identically satisfied. Substituting, in turn, into the equations (4.8.4), we get the following equations

$$\nabla_1^2\nabla_1^2 F + 2\mu m\nabla_1^2\vartheta = 0\,, \tag{4.8.7}$$

$$\nabla_1^2(l^2\nabla^2 - 1)\Psi = 0\,, \tag{4.8.8}$$

where

$$l^2 = \frac{(\gamma+\varepsilon)(\mu+\alpha)}{4\mu\alpha}\,, \qquad m = \frac{(3\lambda+2\mu)}{\lambda+2\mu}\alpha_t\,.$$

The functions F and Ψ are not mutually independent. There are connected by the relations

$$\partial_1\gamma_{21} - \partial_2\gamma_{11} - \varkappa_{13} = 0\,, \qquad \partial_1\gamma_{22} - \partial_2\gamma_{12} - \varkappa_{23} = 0\,,$$

$$\partial_1\varkappa_{23} - \partial_2\varkappa_{13} = 0\,. \tag{4.8.9}$$

Consequently, we obtain

$$\begin{aligned} -\partial_1(1 - l^2\nabla_1^2)\Psi &= A\partial_2\nabla_1^2 F + B\partial_2\vartheta\,, \\ \partial_2(1 - l\,\nabla_1^2)\Psi &= A\partial_1\nabla_1^2 F + B\partial_1\vartheta\,, \end{aligned} \tag{4.8.10}$$

where

$$A = \frac{(\lambda+2\mu)(\gamma+\varepsilon)}{4\mu(\lambda+\mu)}\,, \qquad B = \frac{\nu(\gamma+\varepsilon)}{2(\lambda+\mu)}\,.$$

We have still to give the boundary conditions for Eqs.(4.8.7), (4.8.8). We assume the boundary s to be free of loadings. This condition is expressed by the equations

$$\sigma_{ji}n_j = 0\,, \quad \mu_{j3} = 0\,, \quad j = 1,2\,, \quad \underline{x} \in s\,, \tag{4.8.11}$$

which, if expressed in Ψ and F , lead to the following ones

$$\frac{d}{ds}(\partial_2 F - \partial_1 \Psi) = 0 \ , \quad \frac{d}{ds}(\partial_1 F + \partial_2 \Psi) = 0 \ , \qquad \frac{\partial \Psi}{\partial n} = 0 \ . \tag{4.8.12}$$

The quantities $\frac{\partial}{\partial s}$ and $\frac{\partial}{\partial n}$ are the derivatives along the boundary S and along the normal to this boundary.

The solutions of Eqs.(4.8.7) may be written in the following form

$$F = F' + F'' \ , \qquad \Psi = \Psi' + \Psi'' . \tag{4.8.13}$$

Here F' is the particular integral of the equation

$$\nabla_1^2 F' + 2\mu m \theta = 0 \ , \qquad \Psi' = 0 \ . \tag{4.8.14}$$

It may easily be checked that the functions F' and Ψ' lead to the symmetric stress tensor

$$\sigma'_{ji} = \sigma'_{ij} = -\partial_j \partial_i F' + \delta_{ji} \nabla_1^2 F' \ , \qquad \mu'_{ji} = 0 \ . \tag{4.8.15}$$

The stress function F' is the particular integral for Hooke's medium and, at the same time, the particular solution of the equation in classical thermoelasticity.

Finally, the conditions (4.8.10) are satisfied. They take here the form:

$$\partial_1(\nabla_1^2 F' + 2\mu m \theta) = 0 \qquad \partial_2(\nabla_1^2 F' + 2\mu m \theta) = 0 \tag{4.8.16}$$

leading to Eq.(4.8.14).

The functions F'', Ψ'' have to verify the following equations

$$\nabla_1^2\nabla_1^2 F'' = 0 \quad , \qquad \nabla_1^2(1 - l^2\nabla_1^2)\Psi'' = 0 , \tag{4.8.17}$$

with boundary conditions

$$(\sigma'_{ji} + \sigma''_{ji})n_j = 0 , \; \mu''_{3j}n_j = 0 , \; i,j = 1,2 \; . \tag{4.8.18}$$

At the same time, the following relations have to be satisfied

$$\begin{aligned} -\partial_1(1 - l^2\nabla_1^2)\Psi'' &= A\partial_2\nabla_1^2 F'' , \\ \partial_2(1 - l^2\nabla_1^2)\Psi'' &= A\partial_1\nabla_1^2 F'' . \end{aligned} \tag{4.8.19}$$

Thus, we succeeded in proving, in general, that for the plane state of strain the solution of the thermoelastic problem may consist of two parts:

a) The particular solution of non-homogeneous equations of classical thermoelasticity and

b) The general solution of homogeneous differential equations of micropolar elasticity (with $\vartheta = 0$).

Let us return once more to the system (4.8.7) (4.8.8) with the homogeneous boundary conditions (4.8.12). Consider the state of stress in an infinitely long cylinder heated on the side surface. Assuming the absence of sources, the temperature satisfies the Laplace equation. In this case (4.8.7) (4.8.8) and the boundary conditions are homogeneous.

The solution of the system is trivial, that is to

say

$$F = 0 \ , \quad \Psi = 0$$

only when the temperature ϑ is constant. Only under this assumption are the additional relations (4.8.10) satisfied.

Thus, we find that for $\vartheta =$ const. the stress functions F and Ψ in a simply-connected cylinder vanish, which leads to the zero values of σ_{11} , σ_{22} , σ_{12} , σ_{21} , μ_{13} , μ_{31} , μ_{32} , μ_{32} . The only non-vanishing stress is σ_{33} , given by the formula

$$\sigma_{33} = \lambda \gamma_{kk} - \nu \vartheta = - \frac{\mu(3\lambda + 2\mu)}{\lambda + \mu} \alpha_t \ . \qquad (4.8.20)$$

All other temperature fields satisfying the Laplace equation lead to non-vanishing stresses.

Consider the elastic semi-space $x_1 \geq 0$ heated in the plane $x_1 = 0$ to the temperature $f(x_2)$. We assume that this plane is free of stress and hence, for $x_1 = 0$ we have

$$\sigma_{11} = 0 \ , \quad \sigma_{12} = 0 \ , \quad \mu_{13} = 0 \ . \qquad (4.8.21)$$

To determine the stresses we apply the exponential Fourier transform

$$\tilde{g}(x_1, \zeta) = \frac{1}{\sqrt{2\pi}} \int_{-\infty}^{\infty} g(x_1, x_2) e^{i\zeta x_2} dx_2$$
$$g(x_1, x_2) = \frac{1}{\sqrt{2\pi}} \int_{-\infty}^{\infty} \tilde{g}(x_1, \zeta) e^{-i\zeta x_2} d\zeta \ . \qquad (4.8.22)$$

First, we determine the temperature $\vartheta(x_1, x_2)$ by solving the Laplace equation

$$\nabla_1^2 \vartheta = 0 \tag{4.8.23}$$

with the boundary condition

$$\vartheta(0, x_2) = f(x_2) \tag{4.8.24}$$

and the regularity condition at infinity. By making use of the Fourier transform, we express the temperature in the form of the integral

$$\vartheta(x_1, x_2) = \frac{1}{\sqrt{2\pi}} \int_{-\infty}^{\infty} \tilde{f}(\zeta) e^{-\zeta x_1 - i x_2 \zeta} d\zeta , \tag{4.8.25}$$

where

$$\tilde{f}(\zeta) = \frac{1}{\sqrt{2\pi}} \int_{-\infty}^{\infty} f(x_2) e^{i\zeta x_2} d\zeta .$$

To determine the state of stress, we introduce the functions F,, Ψ considered in the preceding section. Set

$$F = F' + F'', \quad \Psi = \Psi' + \Psi'' \tag{4.8.26}$$

where F' , Ψ', are particular solutions of the system of Eqs. (4.8.7) (4.8.8).

Let us assume that $\Psi' = 0$ and that the function satisfies the differential equation

$$\nabla_1^2 F' + 2\mu m \vartheta = 0 \tag{4.8.27}$$

with the boundary condition

$$F' = 0 \quad \text{for} \quad x_1 = 0 \tag{4.8.28}$$

and the regularity condition for $|x_1^2 + x_2^2| \to \infty$.

By applying the exponential Fourier transform, we have

$$F'(x_1, x_2) = \frac{\mu m x_1}{\sqrt{2\pi}} \int_{-\infty}^{\infty} \frac{f(\zeta)}{\zeta} e^{-\zeta x_1 - i\zeta x_2} d\zeta \; . \tag{4.8.29}$$

From the formulae (4.8.15) we determine the stresses connected with the function F':

$$\begin{aligned}
\sigma'_{11} &= -\frac{\mu m x_1}{\sqrt{2\pi}} \int_{-\infty}^{\infty} \zeta \tilde{f}(\zeta)\, e^{-\zeta x_1 - i\zeta x_2} d\zeta \, , \\
\sigma'_{22} &= -\frac{\mu m}{\sqrt{2\pi}} \int_{-\infty}^{\infty} \tilde{f}(\zeta)(2 - x_1\zeta) e^{-\zeta x_1 - i\zeta x_2} d\zeta \, , \\
\sigma'_{12} = \sigma'_{21} &= \frac{i\mu m}{\sqrt{2\pi}} \int_{-\infty}^{\infty} \tilde{f}(\zeta)(1 - x_1\zeta) e^{-\zeta x_1 - i\zeta x_2} d\zeta \, , \\
\mu'_{13} = \mu'_{23} &= 0 \; .
\end{aligned} \tag{4.8.30}$$

The functions F, Ψ should satisfy the equations

$$\nabla_1^2 \nabla_1^2 F'' = 0 \quad , \qquad \nabla_1^2 (1 - l^2 \nabla_1^2) \Psi'' = 0 \; , \tag{4.8.31}$$

with the boundary conditions

$$\sigma'_{11} + \sigma'_{22} = 0 \; , \; \sigma'_{12} + \sigma''_{12} = 0 \; , \; \mu''_{13} = 0 \qquad \text{for} \qquad x_1 = 0 \; . \tag{4.8.32}$$

By applying the exponential Fourier transform to Eqs.(4.8.31) and taking into account the regularity conditions at infinity, we arrive at the integrals

(4.8.33') $$F''(x_1,x_2)=\frac{1}{\sqrt{2\pi}}\int_{-\infty}^{\infty}(M+N\zeta x_1)e^{-\zeta x_1-i\zeta x_2}d\zeta ,$$

(4.8.33") $$\Psi''(x_1,x_2)=\frac{1}{\sqrt{2\pi}}\int_{-\infty}^{\infty}(Ce^{-\zeta x_1}+De^{-\varrho x_1})e^{-i\zeta x_2}d\zeta ,$$

$$\varrho=\left(\zeta^2+\frac{1}{l^2}\right)^{1/2} .$$

The quantities M, N, C and D are to be determined by means of the boundary conditions (4.8.32) and the relations (4.8.10). The boundary conditions (4.8.32) in terms of the functions F'', Ψ'' take the form

(4.8.34) $$|\sigma'_{11}+\partial_2^2F''-\partial_1\partial_2\Psi''|_{x_1=0}=0 , \quad |\sigma'_{12}-\partial_1\partial_2F''-\partial_2^2\Psi''|_{x_1=0}=0 ,$$
$$|\partial_1\Psi''|_{x_1=0}=0 .$$

The first boundary condition leads to the result $M=0$. The last yields the relation

$$\zeta C+\varrho D=0 .$$

Finally, the second condition (4.8.34) for the Fourier transform has the form

$$|\tilde{\sigma}'_{12}+\tilde{\sigma}''_{12}|_{x_1=0}=|i\mu m\tilde{f}(1-\zeta x_1)e^{-\zeta x_1}+i\zeta\tilde{F}''+\zeta^2\tilde{\Psi}''|_{x_1=0}=0 ,$$

and leads to the relation

(4.8.35) $$i\mu m\tilde{f}+iN\zeta^2+C\zeta^2\left(1-\frac{\zeta}{\varrho}\right)=0 .$$

We still have to satisfy Eqs.(4.8.10). In the transformed form they are

$$\begin{aligned} l^2\partial_1(\partial_1^2-\rho^2)\tilde{\Psi}'' &= -i\zeta A(\partial_1^2-\zeta^2)\tilde{F}'',\\ -i\zeta l^2(\partial_1^2-\rho^2)\tilde{\Psi}'' &= A\partial_1(\partial_1^2-\zeta^2)\tilde{F}''. \end{aligned} \tag{4.8.36}$$

By introducing $\tilde{\Psi}''$, $\tilde{F}''$ into these equations, we obtain the relation

$$C = 2iN\zeta^2 A \ .$$

Thus, we determine the quantities C , D , M and N :

$$N = -\frac{\mu m\tilde{f}}{\zeta^2\Delta_0}, \qquad C = -\frac{2Ai\mu m\tilde{f}}{\Delta_0}, \qquad D = -\zeta\frac{C}{\rho}, \qquad M = 0 .$$

Here

$$\Delta_0 = 1 + 2A\zeta^2\left(1-\frac{\zeta}{\rho}\right) .$$

Hence we obtain the stress functions F'', Ψ'' in terms of the Fourier integrals

$$F'' = -\frac{\mu m x_1}{\sqrt{2\pi}}\int_{-\infty}^{\infty}\frac{\tilde{f}(\zeta)}{\zeta\Delta_0}e^{-\zeta x_1 - i x_2\zeta}\,d\zeta , \tag{4.8.37}$$

$$\Psi'' = -\frac{2i\mu m A}{\sqrt{2\pi}}\int_{-\infty}^{\infty}\frac{\tilde{f}(\zeta)}{\Delta_0}\left(e^{-\zeta x_1}-\frac{\zeta}{\rho}e^{-\rho x_1}\right)e^{-i x_2\zeta}\,d\zeta . \tag{4.8.38}$$

We still have to find the total stresses by means of the formulae (4.8.15). We have here

$$\sigma_{11} = -\frac{\mu m}{\sqrt{2\pi}}\int_{-\infty}^{\infty}\tilde{f}(\zeta)\left[x_1\zeta\left(1-\frac{1}{\Delta_0}\right)e^{-\zeta x_1}+\frac{2A\zeta^2}{\Delta_0}\left(e^{-\zeta x_1}-e^{-\rho x_1}\right)\right]e^{-i\zeta x_2}\,d\zeta, \tag{4.8.39}$$

$$\sigma_{22} = -\frac{\mu m}{\sqrt{2\pi}} \int_{-\infty}^{\infty} \tilde{f}(\zeta)\left[(2 - x_1\zeta)\left(1 - \frac{1}{\Delta_0}\right)e^{-\zeta x_1} - \frac{2A\zeta^2}{\Delta_0}(e^{-\zeta x_1} - e^{-\rho x_1})\right] e^{-i x_2 \zeta} d\zeta ,$$

$$\sigma_{12} = \frac{i\mu m}{\sqrt{2\pi}} \int_{-\infty}^{\infty} \tilde{f}(\zeta)\left[(1 - \zeta x_1)\left(1 - \frac{1}{\Delta_0}\right)e^{-\zeta x_1} - \right.$$

$$\left. - \frac{2A\zeta^2}{\Delta_0}\left(e^{-\zeta x_1} - \frac{\zeta}{\rho} e^{-\rho x_1}\right)\right] e^{-i\zeta x_2} d\zeta ,$$

$$\sigma_{21} = \frac{i\mu m}{\sqrt{2\pi}} \int_{-\infty}^{\infty} \tilde{f}(\zeta)\left[(1 - \zeta x_1)\left(1 - \frac{1}{\Delta_0}\right)e^{-\zeta x_1} + \right.$$

$$\left. + \frac{2A\zeta^2}{\Delta_0}(\zeta e^{-\zeta x_1} - \rho e^{-\rho x_1})\right] e^{-i\zeta x_2} d\zeta$$

$$\mu_{13} = \frac{2i\mu m A}{\sqrt{2\pi}} \int_{-\infty}^{\infty} \frac{\tilde{f}(\zeta)}{\Delta_0} \zeta (e^{-\zeta x_1} - e^{-\rho x_1}) e^{-i\zeta x_2} d\zeta \quad ,$$

$$\mu_{23} = -\frac{2i\mu m A}{\sqrt{2\pi}} \int_{-\infty}^{\infty} \frac{\tilde{f}(\zeta)}{\Delta_0}\left(e^{-\zeta x_1} - \frac{\zeta}{\rho} e^{-\rho x_1}\right) e^{-i\zeta x_2} d\zeta .$$

Consider now the normal stress

$$\sigma_{33} = \gamma_{kk}\lambda - \nu\theta = \frac{\lambda}{2(\lambda + \mu)}(\sigma_{11} + \sigma_{22}) - \frac{\mu\nu\theta}{\lambda + \mu} .$$

We have

$$\sigma_{33} = -\frac{\mu m \lambda}{(\lambda + \mu)\sqrt{2\pi}} \int_{-\infty}^{\infty} \tilde{f}(\zeta)\left(1 - \frac{1}{\Delta_0}\right) e^{-\zeta x_1 - i\zeta x_2} d\zeta - \frac{\nu\mu\theta}{\lambda + \mu} . \tag{4.8.40}$$

In the particular case of the Hookean body we set in the stresses $\alpha = 0$ (whence $l^2 \to 0$, $\rho \to \zeta$, $\Delta_0 \to 1$). It is readily observed that in this case the stresses σ_{11}, σ_{12}, σ_{21}, σ_{22}, μ_{13}, μ_{23}, μ_{31}, μ_{32} vanish. Only the stress σ_{33} remains different from zero. These results agree with the well known I.N. Muskhelishvili theorem[+)].

For the stress σ_{33} we have

$$\sigma_{33} = -\frac{\nu\mu}{\lambda+\mu}\,\vartheta(x_1, x_2) \,. \qquad (4.8.41)$$

In the case of the micropolar body the stresses are in principle different from zero, except for the case of constant temperature.

Let us now calculate the stress σ_{22} on the boundary of the semi-space. We have

$$\sigma_{22}(0, x_2) = -\frac{2\mu m}{\sqrt{2\pi}}\int_{-\infty}^{\infty} \bar{f}(\zeta)\left(1 - \frac{1}{\Delta_0}\right)e^{-i\zeta x_2}\,d\zeta \,. \qquad (4.8.42)$$

By assuming that the quantity $1/l^2$ is very small as compared with unity, we expand ρ into an infinite series. By taking only two terms of this expansion, we obtain

$$\rho = \zeta\left(1 + \frac{1}{\zeta^2 l^2}\right)^{1/2} \approx \zeta\left(1 + \frac{1}{2l^2\zeta^2}\right) \,.$$

+) Muskhelishvili, I.N.: Basic problems of the mathematical theory of elasticity. (in Russian). 3-rd ed., Moscow - Leningrad, 1948.

Thus, we have

$$1 - \frac{1}{\Delta_0} = \sigma \frac{\zeta^2}{\zeta^2 + k^2}$$

where

$$\sigma = \frac{(\lambda + 2\mu)\alpha}{\mu(\lambda + \mu) + \alpha(3\mu + 2\lambda)} \quad , \quad k^2 = \frac{2\mu\alpha(\lambda + \mu)}{(\gamma + \varepsilon)[\alpha(2\lambda + 3\mu) + \mu(\lambda + \mu)]} .$$

Assume that the distribution of temperature on the boundary $x_1 = 0$ is given by the expression

$$\theta(0, x_2) = \theta_0 [H(x_2 - c) - H(x_2 + c)] \ , \tag{4.8.43}$$

where $H(x)$ is the Heaviside function. Over the infinite strip $|x_2| \le c, -\infty < x_3 < \infty$ there acts the temperature θ_0, while for $|x_2| > c, -\infty < x_3 < \infty$ we have $\theta = 0$. The formula (4.8.42) takes the form

$$\sigma_{22}(0, x_2) = - \frac{4\mu m \sigma \theta_0}{\pi} \int_0^\infty \frac{\zeta \sin \zeta c}{\zeta^2 + k^2} \cos \zeta x_2 d\zeta \ .$$

This integral can be solved. Namely, we obtain

$$\sigma_{22}(0, x_2) = 2\mu m \sigma \theta_0 \begin{cases} -e^{-kc} \operatorname{ch}(k x_2) & x_2 < c \\ \\ e^{-k|x_2|} \operatorname{sh}(kc) & x_2 > c \ . \end{cases} \tag{4.8.44}$$

Bibliography.

[1] Achenbach, J.D.: Free vibration of a layer of micropolar continuum. Int.J.Engng.Sci. 7, 10 (1969), 1025.

[2] Adomeit, G.: Ausbreitung elastischer Wellen in einen Cosserat-Kontinuum mit freiner Oberfläche. ZAMM. 46, Sonderheft (1966), 158.

[3] Adomeit, G.: Determination of elastic constants of a structured material, IUTAM - Symposium, 1967, Freudenstadt - Stuttgart, Mechanics of generalized continua, (1968), 80.

[4] Ablas, J.B.: The Cosserat continuum with electronic spin, IUTAM - Symposium, 1967, Freudenstadt - Stuttgart, Mechanics of generalized continua (1968), 350.

[5] Ariman, T.: Some problems in bending of micropolar plates, Bull.Acad.Polon.Sci.Ser.Sci.Techn. I - 16, 7 (1968), 295; II - 16, 7 (1968), 301.

[6] Ariman, T.: Micropolar and dipolar fluids. Int.J.Engng. Sci. 6, 1. (1968), 1.

[7] Askar, A. and Cakmak A.S.: A structural model of a micro polar continuum. Int.J.Engng.Sci. 6, 10 (1968), 583.

[8] Atsumi A. and S.Itou,: Waves produced in the elastic Cosserat half-plane by conducting heat and moving load. Arch.Mech.Stos. 12, 1, (1970), 75.

[9] Baranski, W. and Wozniak Cz.: The fibrous body as a sim ply connected model of multi-hole disc,Arch.Mech. Stos. 18, 3 (19 6), 274.

[10] Bogy, D.B. and E. Sternberg,: The effect of couple-

-stresses on the corner singularity due an a-symmetric shear loading, Int.J.Solids Structures, 4, (1968), 159.

[11] Bresson,A.: Sui sistemi continui nel caso asimmetrico, Ann. di Mat. Pura e Applicata, S. IV. 62 (1963).

[12] Carlson, D.E.: Stress functions for couple and dipolar stresses, Quart. Appl. Math. 24, 1, (1966), 29.

[13] Carlson, D.E.: On Günther's stress functions for couple stresses, Quart. Appl. Math. 25, 2 (1967), 139.

[14] Carlson, D.E.: The general solution of the stress equation of equilibrium for a Cosserat continuum. Proc. Fifth. U.S. Nat. Congr. Appl. Mech. (1966), 249.

[15] Mc.Carthy M.F. and A.C. Eringen,: Micropolar viscoelastic waves. Int.J.Engng. Sci. 7, (1969) 447.

[16] Chauhan R.S.: Couple stresses in a curved bar. Int.J. Engng.Sci. 7, 8 (1969), 895.

[17] Cosserat E. and F.Cosserat,: Sur la théorie de l'élasticité. Ann. de l'Ecole Normale de Toulouse, 10, 1 (1896) 1.

[18] Cosserat E. and F.Cosserat,: Theory des corps déformables, A.Herman et Fils, Paris, 1909.

[19] Crochet M.J. and P.M.Naghdi: Large deformation solutions for an elastic Cosserat surface, Int.J.Engng.Sci. 7, 3 (1969), 309.

[20] Doyle,J.N.: Singular solutions in elasticity, Acta Mechanica, 1966.

[21] Ejke, B.C.O.: The plane circular cracks problem in the linearized couple-stress theory, 7, 9 (1969), 947.

[22] Eringen, A.C.: Mechanics of micromorphic continua, IUTAM Symposium, 1967, Freudenstadt - Stuttgart, Mechan

ics of generalized continua (1968), 18.

[23] Eringen, A.C. and E.S. Suhubi: Nonlinear theory of simple microelastic solids. Int.J.Engng.Sci., I, 2, 2 (1964), 189; II, 2, 4 (1964), 389.

[24] Eringen, A.C.: Linear theory of micropolar elasticity, J.Math.Mech. 15 (1966),909.

[25] Eringen, A.C.: Theory of micropolar plates. ZAMP 18, (1967), 567.

[26] Eringen, A.C.: Mechanics of micropolar materials. Proc. of the XIth Intern. Congr. of Applied Mechanics, München, Springer (1965), 131.

[27] Eringen, A.C.: Theory of micropolar elasticity. Fracture, Vol.II, (1968), Academic Press, New York.

[28] Eringen, A.C.: Linear theory of micropolar viscoelasticity. Int.J.Engng.Sci. 5, (1967), 191.

[29] Graff, K.F.and Yih-Hsing Pao,: The effects of couple-stresses on the propagation and reflection of plane waves in an elastic half-space, J.of Sound and Vibration, 6, 2 (1967), 217.

[30] Green,A.E.: Micro-materials and multipolar continuum mechanics, Int.J.Engng.Sci., 3, 5 (1965), 533.

[31] Green, A.E. and P.M.Naghdi: The Cosserat surface. IUTAM Symposium, 1967, Freudenstadt - Stuttgart, Mechanics of generalized continua (1968), 36.

[32] Green, A.E.,P.M. Naghdi, and W.L. Wainright: A general theory of Cosserat surface, Arch.Rat.Mech.Analysis, 20 (1965), 287.

[33] Green, A.E., R.S.Rivlin: Multipolar continuum mechanics, Arch.Rat.Mech.Anal., 17, 2 (1964), 113.

[34] Grioli, G.: Mathematical Theory of Elastic Equilibrium (Recent Results), Ergebnisse der Angewandten Ma

thematik. 7, (1962), 141.

[35] Grioli, G.: Sulla meccanica dei continui a trasformazioni reversibili. Seminari dell'Istituto Nazionale di Alta Matematica, 1962-63.

[36] Grioli, G.: On the thermodynamic potential of Cosserat Continua, Symposium IUTAM Freudenstadt - Stuttgart (1967).

[37] Grioli, G.: On the thermodynamic potential for continuums with reversible transformations - some possible types, Meccanica 1, No 1/2, 1966, 15.

[38] Grioli, G.: Questioni di compatibilità per i continui di Cosserat, Ist. Nazionale di Alta Matematica. Symposia Matematica, 1 (1968), 272.

[39] Grioli, G.: Elasticità asimmetrica, Annali di Mat. Pura e Applicata, 4, 4 (1960), 389.

[40] Grioli, G.: Onde di discontinuità ed asimmetriche, Acc. Naz. dei Lincei, S. VIII, V. XXIX, fasc. 5, Nov. 1960.

[41] Grot, R.A.: Thermodynamics of a continuum with microstructure, Int.J.Engng.Sci. 7, 8 (1969), 801.

[42] Günther, W.: Zur Statik und Kinematik des Cosseratschen Kontinuums, Abh. der Braunschw. Wiss. Geselschaft, 10 (1958), 195.

[43] Günther, W.: Analoge Systeme von Schalengleichungen, Ing.- Archiv. 30 (1961), 160.

[44] Hartranft, R.J. and G.C. Sih: The effect of couple - - stresses on the stress concentrations of a circular inclusion, Trans. ASME, ser. E.J. Appl. Mech. 32 (1965), 429.

[45] Hermann, G. and J.D. Achenbach: Applications of theories of generalized Cosserat continua to the dynamics

of composite materials, IUTAM - Symposium, 1967, Freudenstadt - Stuttgart, Mechanics of generalized continua (1968), 69.

[46] Hlavaček,I.: On the existence and uniqueness of solution and some variational principles in linear theories with couple-stresses, I. Cosserat continuum, II, Mindlin's elasticity with microstructure on the first strain-gradient theory. Aplikace Matematiky, 14, 5, (1969), 387 - 427.

[47] Hofman,O., F.O.F. Shahman,: Physical model of a 3-constant isotropic elastic material, Trans. ASME, ser.E.J. Appl.Mech., 32, 4, (1965), 837.

[48] Huilgol,R.R.: On the concentrated force problem for two--dimensional elasticity with couple-stresses, Int. J.Engng.Sci. 5, 1,(1967), 81.

[49] Iesan,D.: On the linear theory of micropolar elasticity. Int.J.Engng.Sci. I. 12, (1969), 1213.

[50] Iesan,D.: On the plane coupled micropolar thermoelasticity, Bull.Acad.Polon.Sci.,Sér.Sci.Tech. I - 16, 8, (1968), 379; II - 16, 8,(1968), 385.

[51] Kaliski,S.: On a model of the continuum with essentially non-symmetric tensor of mechanical stress, Arch. Mech. Stos. 15, (1963).

[52] Kaliski,S.: Thermo-magneto-microelasticity, Bull.Polon. Sci.Sér.Sci.Techn. 16, 1, (1968),7.

[53] Kaliski,S.,J.Kapelewski and C.Rymarz: Surface waves on an optical branch in a continuum with rotational degrees of freedom, Proc.Vibr.Probl. 9, 2,(1968), 108.

[54] Kaloni,P.N. and T.Ariman: Stress concentration effects in micropolar elasticity, ZAMP, 18, 1, (1967),136.

[55] Kessel,S.:Lineare Elastizitätstheorie des anisotropen

Cosserat-Kontinuums, Abh. derBraunschw. Wiss. Gesellschaft, 16, (1964), 1.

[56] Kessel,S.: Die Spannungsfunktionen des Cosserat-Kontinuums ZAMM, 47, 5, (1967), 329.

[57] Kessel,S.: Stress functions and loading singularities for the infinitely extended, linear elastic - isotropic Cosserat continuum, IUTAM - Symposium, 1967, Freudenstadt - Stuttgart, Mechanics of generalized continua, (1968), 114.

[58] Klemm,P.,Cz.Wozniak,: Perforated circular plate under large deflection,Arch.Mech.Stos. 19, 1,(1969),45.

[59] Koiter,W.T.: Couple-stresses in the theory of elasticity. Koninkl. Nederl. Akad. van Wetenschappen. Proc., ser.B. I, 67, 1, (1964), 17; II, 67, 1, (1964), 30.

[60] Kröner,E.: Das physikalische Problem der antisymmetrischen Spannungen und der sogenannten Momentenspannungen, Appl. Mech. Proc. XIth Int. Congress Appl. Mech., München, 1964 (1966), 143.

[61] Kröner,E.: Interrelations between various branches of continuum mechanics, IUTAM - Symposium, 1967, Freudenstadt - Stuttgart, Mechanics of generalized continua, (1968), 330.

[62] Kunin,I.A.: The theory of elastic media with microstructure and theory of dislocations, IUTAM - Symposium, 1967, Freudenstadt - Stuttgart, Mechanics of generalized continua (1968), 321.

[63] Kuvshinskii,E.V. and E.L.Aero: Continuum theory of asymmetric elasticity (in Russian), Fizika Twerdego Tela 5 (1963), 2592.

[64] Mindlin, R.D.: Influence of couple-stresses on stress - concentrations, Exper. Mech., 3, 1 (1963), 1.

[65] Mindlin,R.D.: Microstructure in linear elasticity, Arch. Rat.Mech.Analysis, 16, 1 (1963), 51.

[66] Mindlin,R.D.: Representation of displacements and stresses in plane strain with couple-stresses, IUTAM - Symposium, 1963, Tbilisi (1964), 256.

[67] Mindlin,R.D.: On the equations of elastic materials with microstructure, Int.J.Solids Struct., 1, 1,(1965), 73.

[68] Mindlin,R.D.: Stress functions for a Cosserat continuum, Int.J.Solids Struct., 3 (1965), 417.

[69] Mindlin,R.D.: Theories of elastic continuum and crystal lattice theories, IUTAM - Symposium, 1967, Freudenstadt - Stuttgart, Mechanics of generalized continua (1968), 312.

[70] Mindlin,R.D.; H.F.Tiersten: Effects of couple-stresses in linear elasticity, Arch. Rat. Mech. Anal., 11, 5,(1962), 415.

[71] Misicu,M.: A generalization of the Cosserat equations for the motion of deformable bodíes (with internal degrees of freedom), Rev. Roum. Sci. Techn., Sér. de Méc. Appl., 9, (1964), 1351.

[72] Muki,R. and E. Sternberg: The influence of couple--stresses on singular stress contcentrations in elastic solids. Z.A.M.P. 16, 5 (1965), 611.

[73] Naghdi,P.M.: On a variational theorem in elasticity and its application to shell theory, Trans. ASME, ser. E.J.Appl.Mech., 31, 4, (1964), 647.

[74] Naghdi,P.M.: A static-geometric analogue in the theory of couple-stresses, Proc.Konincl.Nederlandse Akad.Wet.,ser.B. 68 (1965),29.

[75] Neuber,H.: On the general solution of linear-elastic problem in anisotropic and isotropic Cosserat

continua, Appl.Mech.Proc. XIth Int. Congress Appl. Mech., München 1964, (1966), 153.

[76] Neuber,H.: Über Probleme der Spannungskonzentration im Cosserat-Körper. Acta Mechanika, 2, 1, (1966),48.

[77] Neuber,H.: Die schubbeauspruchte Kerbe im Cosserat-Korper, Z.A.M.M. 47, 5, (1967),313.

[78] Neuber,H.: On the effects of stress concentration in Cosserat continuum, IUTAM - Symposium, 1967, Freudenstadt - Stuttgart, Mechanics of generalized continua (1968), 109.

[79] Nowacki,W.: Couple-stresses in the theory of thermoelasticity. I. Bull. Acad. Polon. Sci. Sér. Sci. Tech. 14 (1966), 97; II - 14 (1966), 293; III - 14 (1966), 505.

[80] Nowacki,W.: On the completeness of stress functions in asymmetric elasticity. Bull. Acad. Polon. Sci. Sér. Sci. Techn. 16, 7,(1968), 309.

[81] Nowacki, W.: Propagation of rotation waves in asymmetric elasticity. Bull. Acad. Polon. Sci. Sér. Sci. Techn. 16, 10, (1968), 493.

[82] Nowacki, W.: Some theorems of asymmetric thermoelasticity. Bull. Acad. Polon. Sci. Sér. Sci. Techn. 15, 5, (1967), 289.

[83] Nowacki, W.: Couple-stresses in the theory of thermoelasticity, Proc. of IUTAM - Symposium on Irreversible Aspects in Continuum Mechanics, Vienna 1966, J. Springer (1968), 259.

[84] Nowacki, W.: On the completeness of potentials in micropolar elasticity. Arch. Mech. Stos. 21, 2, (1969), 107.

[85] Nowacki, W.: Green functions for micropolar elasticity. Proc. Vibr. Probl. 10, 1, (1969), 3.

[86] Nowacki,W. and W.K.Nowacki: Generation of waves in an infinite micropolar elastic solid, Bull. Acad. Polon. Sci. Sér. Sci. Techn. I - 17, 2, (1969), 75; II - 17, 2, (1969), 83.

[87] Nowacki, W.: On certain thermoelastic problems in micro polar elasticity. Bull. Acad. Polon. Sci. Sér. Sci. Techn. 17, 5, (1969), 273.

[88] Nowacki, W.: Formulae for overall thermoelastic deforma tion in a micropolar body. Bull. Acad. Polon.Sci. Sér. Sci. Techn. 17, 1, (1969), 257.

[89] Nowacki,W. and W.K.Nowacki: The plane Lamb problem in a semi-infinite micropolar elastic body. Arch. Mech. Stos. 21, 3, (1969), 241.

[90] Nowacki,W. and W.K.Nowacki: The generation of waves in an inifinite micropolar elastic solid. Proc.Vibr. Probl. 10, 2 (1969), 170.

[91] Nowacki,W. and W.K.Nowacki: The axially symmetrical Lambs problem in a semi-infinite micropolar elastic solid. Proc. Vibr. Probl. 10, 2, (1969), 97.

[92] Nowacki,W.: Generalized Love's functions in micropolar elasticity. Bull. Acad. Polon. Sci. Sér. Sci. Techn. 17, 4, (1969), 247.

[93] Nowacki,W. and W.K.Nowacki: Propagation of monochromatic waves in an infinite micropolar elastic plate. Bull. Acad. Polon. Sci. Sér. Sci. Techn. 17, 1, (1969), 29.

[94] Nowacki,W. and W.K.Nowacki: Propagation of elastic waves in a micropolar cylinder. Bull. Acad. Polon. Sci. Sér. Sci. Techn. I - 17, 1, (1969), 39; II - 17, 1, (1969), 49.

[95] Nowacki, W.: Green functions for micropolar elasticity. Bull. Acad. Polon. Sci. Sér. Sci. Techn. 16,

11 - 12, (1968), 555.

[96] Nowacki, W.: Green functions for micropolar thermoelastic. Bull. Acad. Polon. Sci. Sér. Sci. Techn. 16, 11 - 12, (1968), 565.

[97] Nowacki, W.: The plane problem of micropolar thermoelasticity. Arch. Mech. Stos. 22, 1, (1970), 3.

[98] Parfitt, V.R. and A.C. Eringen: Reflection of plane waves from the flat boundary of a micropolar elastic halfspace, Report No 8 - 3, General Technology Corporation.

[99] Palmov, N.A.: Fundamental equations of the theory of asymmetric elasticity. Prikl. Mat. Mekh. 28, (1964), 401. (in Russian).

[100] Palmov, N.A.: Plane problems of the theory of asymmetric elasticity. Prikl. Mat. Mekh. 28, (1964), 1117.

[101] Pietras, F. and J. Wyrwinski: Thermal stresses in a plane anisotropic Cosserat continuum, Arch. Mech. Stos. 19, 5, (1967), 627.

[102] Rajagonal, E.S.: The existence of interfacial couples in infinitesimal elasticity, Ann. der Physik, 6, (1960), 192.

[103] Reissner, E. and F.Y. Wan: A note on Günther's analysis of couple stress, IUTAM - Symposium, 1967, Freudenstadt - Stuttgart, Mechanics of generalized continua (1968), 83.

[104] Rieder, G.: Die Randbedingungen für den Spannungs-functionentensor an ebenen und gekrümmten belasteten Oberflächen. Osterr. Ing. Arch. 18, (1964), 208.

[105] Rivlin, R.S.: Generalized mechanics of continuous media. IUTAM - Symposium, 1967, Freudenstadt - Stuttgart; Mechanics of generalized continua, (1968), 1.

[106] Rymarz, C.: Surface waves in Cosserats medium, Bull. de l'Acad. Pol. Sci.,Sér. Sci. Tech. 15, 3, (1967), 177.

[107] Sandru, N.: Le théorème de réciprocité du type de Betti dans l'élasticité asymétrique, C. rend. hebd.des séances de l'Acad. des Sci., Paris, 260, 13, (1965), 3565.

[108] Sandru, N.: Le théorème de réciprocité asymétrique (cas dynamique), Atti Accad. Naz. dei Lincei, ser.VIII, Rendiconti, Cl. di Sci. fis.,mat.e nat. 38, 1, (1965), 78.

[109] Sandru, N.: On some problems of the linear theory of the asymmetric elasticity. Int.J.Engng. Sci., 4, 1, (1966), 81.

[110] Savin, G.N.: Fundation of couple-stress of elasticity (in Ukrainian), Kiev, 1965.

[111] Sawczuk, A.: On yielding of Cosserat continuum, Arch. Mech. Stos. 19, 3, (1967), 471.

[112] Schade,K.D.: Cosserat-Fläche und Schalentheorie, Doktor thesis, Darmstadt, Tech. Hochschule, (1966).

[113] Schaefer, H.: Continui di Cosserat, Lezioni e Conferenze dell'Università di Trieste, Istituto di Meccanica, Sac. 7 (1965).

[114] Schaefer, H.: Die Spannungsfunktionen des dreidimensionalen Kontinuums und des elastischen Körpers,ZAMM, 33, (1953), 356.

[115] Schaefer, H.: Die Spannungsfunktionen des dreidimensionalen Kontinuums; statische Deutung und Randwerte, Ingenieur - Archiv. 28, (1959), 291.

[116] Schaefer, H.: Versuch einer Elastizitätstheorie des zweidimensionalen ebenen Cosserat-Kontinuums, Miszellaneen der Angew. Mechanik, Festschrift.

W. Tollmien (1962), 277.

[117] Schaefer, H.: Analysis der Motorfelder im Cosserat-Kontinuum, ZAMM, 47, 5 (1967), 319.

[118] Schaefer, H.: Die Spannungsfunktionen eines Kontinuums mit Momentspannungen, Bull. de l'Acad. Pol. Sci., Sér. Sci. Techn., I, 15, 1 (1967), 63; II,- 15, 1 (1967), 69.

[119] Schaefer, H.: Das Cosserat-Kontinuum, ZAMM, 47, 8,(1967), 485.

[120] Schijve, J.: Note on couple-stresses. J.Mech.Solids, 14, 2, (1966), 113.

[121] Smith, A.C.: Inequalities between the constants of a linear microelastic solid, Int.J.Engng.Sci. 6, 2, (1968), 65.

[122] Smith, A.C.: Waves in micropolar elastic. Int.J.Engng. Sci. 5, 10, (1967), 741.

[123] Smith, A.C.: Deformations of micropolar elastic solids, Int.J.Engng.Sci. 5, 8, (1967) 637.

[124] Sokolowski, M.: Couple-stresses in problems of torsion of prismatic bars. Bull. de l'Acad. Polon. Sci. Sér. Sci. Techn. 13, (1965), 419.

[125] Soos, E.: Reciprocity theorems for elastic and thermo - - elastic materials with microstructure, Bull. Acad. Polon. Sci. Sér. Sci. Techn. 16, 11 - 12, (1968), 541.

[126] Soos, E.: Some applications of the theorem on reciprocity for an elastic and thermo-elastic material with micro-structure, Bull. Acad. Polon. Sci.Sér. Sci. Techn. 16, 11 - 12, (1968), 547.

[127] Soos, E.: Uniqueness theorem for homogeneous, isotropic, simple elastic and thermoelastic material having a microstructure, Int.J.Engng.Sci. 7, 3, (1969),

257.

[128] Stefaniak,J.: A generalisation of Galerkin's functions for asymmetric thermoelasticity, Bull.Acad.Polon. Sci.Sér.Sci.Techn. 16, 8, (1968), 391.

[129] Stefaniak,J.: Concentrated loads as body forces, Revue Roumaine de Mathématiques Pures et Appliquées, 14, 4, (1969), 119.

[130] Stefaniak,J.: Reflection of a plane longitudinal wave from a free plane in a Cosserat medium. Arch. Mech. Stos. 21, 6, (1969), 745.

[131] Sternberg, E. and R. Muki: The effect of couple-stresses on the stress concentration around a crack, Int. J.Solids Struct. 3, 1, (1967), 69.

[132] Stojanovic,R. and L.Vujosevic and D.Blagojewic: Couple stresses in thermoelasticity (in print).

[133] Stojanovic,R.: Dislocations in the generalized elastic Cosserat continuum, IUTAM - Symposium, 1967, Freudenstadt - Stuttgart, Mechanics of Generalized Continua, (1968), 152.

[134] Stojanovic, R., S.Djuric and L.Vujosevic: Stress-strain relations for the elastic Cosserat continuum, Mat. Vesnik 1, 16, (1964), 127.

[135] Stojanovic, R. and Vujosevic,L.: Couple-stresses in non-euclidean continua, Publ. de l'Inst. Math. Nouv. Sér. 2, 16, (1962), 71.

[136] Stojanovic,R. and S.Djuric: On the measure of strain in the theory of the elastic generalized Cosserat continuum, Istituto Nazionale di Alta Meccanica, Vol. I, 1968.

[137] Surdia,M.J.: L'action euclidienne de déformation et de mouvement, Mémorial des sciences physiques, Fasc. 29, Paris, 1935.

[138] Suhubi,E.S. and A.C. Eringen: Nonlinear theory of micro elastic solid, Int.J.Engng.Sci. I - 2, 4, (1964); II - 2, 4, (1964), 389.

[139] Tauchert,T.A., W.D.Claus Jr. and T.Ariman: The linear theory of micropolar thermoelasticity, Int.J.Engng. Sci. 6, 1, Engng. (1968), 37.

[140] Teodorescu, P.P.: Sur les corps du type de Cosserat à élasticité linéaire, Istituto Nazionale di Alta Matematica. Symp.Math. 1, (1968), 375.

[141] Teodorescu, P.P.: Sur la notion de moment massique dans le cas des corps du type de Cosserat. Bull. Acad. Polon.Sci.Sér.Sci.Techn. 15, 1, (1967), 57.

[142] Teodorescu, P.P.and N. Sandru: Sur l'action des charges concentrées en elasticité asymétrique plane. Rev. Roumain Math. Pure et Appl. 22, 9 (1967), 1399.

[143] Teodosiu,C.: On the determination of internal stresses in the continuum theory of dislocations, Bull. de l'Acad. Pol. Sci., Sér. Sci. Techn., 12, 12, (1964), 605.

[144] Teodosiu,C.: The determination of stresses and couple--stresses generated by dislocations in isotropic media, Rev.Roum.Sci.Techn., Sér. de Méc. Appl., 10, 6, (1965), 1461.

[145] Teodosiu,C.: Non-linear theory of the materials of grad 2 with initial stresses, I Basic geometric and static equations, Bull. de l'Acad. Pol. Sci., Sér. Sci. Techn., 15, 2, (1967), 103.

[146] Teodosiu,C.: Contributions to the continuum theory of dislocations and initial stresses, Rev. Roum. Sci. Techn., Sér. de Méc. Appl., I - 12,4, (1967),961; II - 12, 5, (1967), 1061; III - 12, 6, (1967), 1291.

[147] Teodosiu,C.: Continuous distribution of dislocations in

hyperelastic material of grade 2., IUTAM - Sympo sium, 1967, Freudenstadt - Stuttgart, Mechanics of generalized continua (1968), 279.

[148] Toupin, R.A.: Elastic materials with couple-stresses, Arch. Rat. Mech. Analysis, 11, 5, (1962), 385.

[149] Toupin, R.A.: Theories of elasticity with couple--stresses, Arch. Rat. Mech. Analysis, 17, 2, (1964), 85.

[150] Toupin, R.A.: Dislocated and oriented media, IUTAM - Sym posium, 1967, Freudenstadt - Stuttgart, Mechanics of generalized continua (1968), 126.

[151] Toupin, R.A.: A dynamical theory of elastic dielectrics, Int.J.Engng.Sci., 1, (1963).

[152] Truesdell,C.: Six lectures on modern natural philisophy, Berlin, 1966, Springer.

[153] Truesdell,C.: Rational mechanics of deformation and flow. Proc. 4-th Int.Congres. Rheol. 1963, 2, (1955), 3.

[154] Truesdell,C. and R.A. Toupin: The classical field theories, Encyclopedia of physics. Vol. III/1. Springer, Berlin, (1960)

[155] Umamaheswaram,S and A.K. Das: Effect of couple stresses on an infinite strip of finite thickness symmetrically loaded. Arch. Mech. Stos. 11, 6, (1969), 753.

[156] Weitsman, Y.: Couple-stresses effects on stress concentrations around a cylindrical inclusion in a field of uniaxial tension, Trans. ASME,ser.E.J. Appl. Mech. 32, (1965), 424.

[157] Wesolowski,Z.: On the couple-stresses in a elastic continuum, Arch. Mech. Stos., 17, (1965), 219.

[158] Wilmanski,K. and C.Wozniak: On geometry of continuous me

dium with microstructure, Arch. Mech. Stos. 19, 5, (1967), 715.

[159] Voigt, W.: Theoretische Studien über die Elastizitätsverhältnisse der Krystalle . I, II, Abh. Königl. Ges. der Wiss., Göttingen, 34, (1887).

[160] Wozniak,Cz.: Theory of fibrous media, Arch. Mech. Stos., I, 17, (1965), 651 - 669; II, 17, (1965), 777.

[161] Wozniak,Cz. and M. Zurawski: On a model of elastic subsoil carrying surface moments, Bull. de l'Acad. Pol. Sér. Sci. Techn., 14, (1966).

[162] Wozniak,Cz.: State of stress in the lattice-type bodies, Bull. Acad. Polon. Sci. Sér. Sci. Techn. 14, (1966), 643.

[163] Wozniak, Cz.: Thermoelasticity of micro-materials, Bull. Acad. Polon. Sci., Sér. Sci. Techn. 14, (1964), 51.

[164] Wozniak, Cz.: Thermoelasticity of the bodies with microstructure, Arch. Mech. Stos. 19, 3, (1967), 335.

[165] Wozniak,Cz.: Thermoelasticity of non simple materials, Arch. Mech. Stos. 19, 4, (1967), 485.

[166] Wozniak, Cz.: Thermoelasticity of non simple continuous media with internal degrees of freedom, Int.J. Engng.Sci. 5, 8, (1967), 605.

[167] Wozniak, Cz.: Load-carrying structures of the dense lattice type. The plane problem. Arch. Mech. Stos. 18, (1966), 581.

[168] Wozniak,Cz.: On the stability of dense plane bar girds, Bull. Acad. Polon. Sci. Sér. Sci. Techn., 13, (1965), 53.

[169] Wyrwinski,J.: Green functions for a thermoelastic Cosserat medium. Bull. de l'Acad. Pol. Sci., Sér.

Sci. Techn., 14, (1966), 145.

[170] Yamamoto,Y.: An intrinsic theory of a Cosserat continuum, RAAG, Research Notes, IIIrd series, No 122.

[171] Zahorski,S.: On motion and thermodynamics of non simple continuum with microstructure. Arch. Mech. Stos. 19, (1967), 25.

Contents

Printed in the United States
By Bookmasters